民航无线电陆空通话(第 2 版)

Radiotelephony Communications
for Air Traffic Controllers and Airline Pilots (2nd Edition)

刘继新 主　编
胡　彬 副主编

国防工业出版社

·北京·

内 容 简 介

《民航无线电陆空通话》以国际民航组织 9432 文件（Manual of Radiotelephony）和 9835 文件（Manual on the Implementation of ICAO Language Proficiency Requirements）为蓝本，以中国民航局空管局最新编修的《空中交通无线电通话用语》为依据编写。本教材包括 9 个单元，共 21 课，内容贯穿实际空中管制与飞行工作的各方面，以国际民航组织和国内的有关文件和资料为背景，以国内外最新使用的各种常规通话用语和非正常情况下的紧急处置通话用语为主要内容，通过大量系统性的句型、对话、术语的学习以及口头和书面的训练，以达到使学员熟练掌握和运用无线电陆空通话用语的目的。

本教材根据我国空中管制和飞行运行的特点，紧扣我国国情，并纳入国际通话程序的最新要求和特点，教材内容具有较强的代表性、时效性和实用性，可作为空管和飞行本科专业教材，也可供广大在职空管和飞行人员参考使用。

图书在版编目（CIP）数据

民航无线电陆空通话 / 刘继新主编；胡彬副主编
. —2 版. —北京：国防工业出版社，2024.4
ISBN 978-7-118-13226-7

Ⅰ.①民… Ⅱ.①刘… ②胡… Ⅲ.①民用航空—陆空协同通信—高等学校—教材 Ⅳ.①V243.1

中国国家版本馆 CIP 数据核字（2024）第 059074 号

※

国防工业出版社出版发行
（北京市海淀区紫竹院南路 23 号 邮政编码 100048）
三河市天利华印刷装订有限公司印刷
新华书店经售

*

开本 787×1092 1/16 印张 11¾ 字数 256 千字
2024 年 4 月第 2 版第 1 次印刷 印数 1—2000 册 定价 58.00 元

（本书如有印装错误，我社负责调换）

国防书店：（010）88540777 书店传真：（010）88540776
发行业务：（010）88540717 发行传真：（010）88540762

前　言

随着我国经济持续稳定的增长，民航业也随之高速发展。党的二十大报告提出，中国应坚持对外开放的基本国策，推动建设开放型世界经济。而作为国民经济的战略性产业，民航具有运输效率高和服务国际化的特征，能够推动实现更大范围、更宽领域、更深层次的对外开放。加强民航从业人员无线电陆空通话能力有助于提高行业服务质量，从而使民航为高质量发展及对外开放事业增添动力。

无线电陆空通话是当今管制员与飞行员信息沟通的主要方式，无线电通话的规范和正确与否直接关系到民航飞行安全。在国际航空史上，由于无线电通话用语不规范、不标准而导致的飞行事故和事故征候屡有发生。近年来，我国航空运输事业持续发展，空中交通流量持续增加，进一步规范无线电陆空通话用语对确保我国空中交通的安全、顺畅具有十分重要的意义。《民航无线电陆空通话（第 2 版）》以国际民航组织的 9432 文件 (Manual of Radiotelephony) 和 9835 文件 (Manual on the Implementation of ICAO Language Proficiency Requirements) 为蓝本，以中国民用航空局空中交通管理局《空中交通无线电通话用语指南》为依据编写，参考了民航局 2023 年最新修订的行业标准《空中交通无线电通话用语》，是一本以提高民航管制学员和飞行学员专业通话技能为主要任务的专业性教材。

本教材包括 9 个单元，共 21 课，内容贯穿实际管制与飞行工作的方方面面，以国际民航组织和国内的有关文件和资料为背景，以国内外最新使用的各种常规通话用语和非正常情况下的紧急处置通话用语为主要内容，通过大量系统性的句型、对话、术语的学习以及口头和书面的训练，以期达到使学员熟练掌握和运用无线电陆空通话用语的目的。本教材配有课文和练习录音，方便读者学习和模仿。读者可以通过扫描教材中各音频文件对应处的二维码播放。

本教材围绕我国管制和飞行实际工作的特点，紧扣我国国情，并纳入国际上陆空通话程序的最新要求和特点，内容具有较强的代表性、时效性和实用性。课后练习可供学员学习和巩固。通过对本教材的学习，可以提高学员的实际通话能力，掌握最新的行业标准，加强学员在处理特殊情况时的信心和应变能力；也可以通过对本教材的学习扩大相关专业的词汇量，提高对飞行各个阶段的理解和掌握，为更好地适应未来的工作做好专业语言上的准备。

　　本教材由南京航空航天大学刘继新副教授主编，胡彬讲师担任副主编。教材编写过程中得到了南京航空航天大学民航学院相关领导和专家的关心和指导，在此深表感谢。

　　由于编者水平的限制，书中难免存在错误和不妥之处，恳请行业专家和读者批评指正。

编　者
2024 年 1 月于南京

Contents

Unit One Basic Operating Procedures 1

  Lesson One Terms, Commonly Used Abbreviations, Numbers, Call Signs, Phonetic
                 Alphabets, Standard Words and Phrases 3
  Lesson Two Establishing Communication and Transfer of Communication 25

Unit Two Preflight to Line-up 29

  Lesson Three ATIS, Arrival/Departure Information, Essential Aerodrome Information,
                 VOLMET 31
  Lesson Four Radio Check and ATC Clearance 37
  Lesson Five Start-up and Push-back 43
  Lesson Six Taxi and Line-up 49

Unit Three Take-off and Departure 55

  Lesson Seven Take-off 57
  Lesson Eight Departure 63

Unit Four En-route 69

  Lesson Nine Level Information and Position Report 71
  Lesson Ten Join, Cross and Leave Airways 77
  Lesson Eleven ADS, RVSM, En-route Holding and Descent 83

Unit Five Approach 91

  Lesson Twelve RNAV, Approach 93
  Lesson Thirteen TCAS, Traffic Circuit 100

Unit Six Landing and After Landing 107

  Lesson Fourteen Missed Approach, Diversion and Local Training 109
  Lesson Fifteen Landing and After Landing, Low Altitude Warning, Terrain Alert ... 115

Unit Seven Radar Control ... 121

Lesson Sixteen Radar Identification and Vectoring, Precision Radar Approach ... 123
Lesson Seventeen Traffic Information and Radar Failure ... 130

Unit Eight Adverse Weather Information ... 135

Lesson Eighteen Icing and Low Visibility ... 137
Lesson Nineteen Turbulence, Thunderstorm and Windshear ... 142

Unit Nine Urgency and Distress ... 147

Lesson Twenty Urgency ... 149
Lesson Twenty-one Distress ... 154

Key to exercises ... 160

参考文献(References) ... 181

Unit One

Basic Operating Procedures

Lesson One

Terms, Commonly Used Abbreviations, Numbers, Call Signs, Phonetic Alphabets, Standard Words and Phrases

Lesson 1

1. Radiotelephony Terms(无线电通话术语)

正切 abeam

An aircraft is "abeam" a fix, point, or object when that fix, point, or object is approximately 90 degrees to the right or left of the aircraft track.

Note: Abeam indicates a general position rather than a precise point.

可用加速-停止距离 accelerate-stop distance available

The length of the take-off run available plus the length of the stopway, if provided.

咨询空域 advisory airspace

An airspace of defined dimensions, or designated route, within which air traffic advisory service is available.

咨询航路 advisory route

A designated route along which air traffic advisory service is available.

机场 aerodrome

A defined area on land or water (including buildings, installations and equipment) intended to be used either wholly or in part for the arrival, departure and surface movement of aircraft.

机场管制服务 aerodrome control service

Air traffic control service for aerodrome traffic.

塔台管制单位 aerodrome control tower

A unit providing air traffic control service for aerodrome traffic.

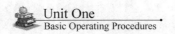

Unit One
Basic Operating Procedures

机场飞行情报服务 aerodrome flight information service

A flight information service provided to aerodrome traffic.

机场活动 aerodrome movement

Aircraft movement in the movement area.

机场运行最低标准 aerodrome operating minima

The limits of usability of an aerodrome for either take-off or landing, usually expressed in terms of visibility, runway visual range, decision altitude, minimum descent altitude and ceiling.

机场交通 aerodrome traffic

All traffic on the maneuvering area of an aerodrome and all aircraft flying in the vicinity of an aerodrome. An aircraft is in the vicinity of an aerodrome when it is in, entering or leaving an aerodrome traffic circuit.

机场起落航线 aerodrome traffic circuit

The specified path to be flown by aircraft operating in the vicinity of an aerodrome.

机场交通地带 aerodrome traffic zone

Airspace of defined dimensions established around an aerodrome for the protection of aerodrome traffic.

航图 aeronautical chart

A representation of a portion of the earth, its culture and relief, specifically designated to meet the requirements of air navigation.

航空移动服务 aeronautical mobile service

A radio communications service between aircraft stations and aeronautical stations, or between aircraft stations.

航空电台 aeronautical station

A land station in the aeronautical mobile service. In certain instances, an aeronautical station may be placed on board a ship or an earth satellite.

机载防撞系统 airborne collision avoidance system

An aircraft system based on SSR transponder signals which operates independently of ground-based equipment to provide advice to the pilot on potential conflicting aircraft that are equipped with SSR transponders.

航空器电台 aircraft station

A mobile station in the aeronautical mobile service on board an aircraft.

航空器识别标志 aircraft identification

A group of letters, figures, or a combination thereof which is either identical to, or coded equivalent of, the aircraft call sign to be used in air-ground communications, and which is used

Lesson One
Terms, Commonly Used Abbreviations, Numbers, Call Signs, Phonetic Alphabets, Standard Words and Phrases

to identify the aircraft in ground-ground air traffic services communications.

航空器危险接近 aircraft proximity

A situation in which, in the opinion of a pilot or controller, the distance between aircraft as well as their relative positions and speed have been such that the safety of the aircraft involved was or may have been compromised.

防空识别区 air defense identification zone

The area of airspace over land or water, extending upward from the surface, within which the ready identification, the location, and the control of aircraft are required in the interest of national security.

机场标高 airport elevation

The elevation of the highest point of the landing area.

空中交通 air traffic

All aircraft in flight or operating on the maneuvering area of an aerodrome.

空中交通管制许可 air traffic control clearance

Authorization for an aircraft to proceed under conditions specified by an air traffic control unit.

Note: (1) For convenience, the term air traffic control clearance is frequently abbreviated to clearance when used in appropriate contexts.

(2) The abbreviated term clearance may be prefixed by the words taxi, take-off, departure, en route, approach or landing to indicate the particular portion of flight to which the air traffic control clearance relates.

空中交通流量管理 air traffic flow management

A service established with the objective of contributing to a safe, orderly and expeditious flow of air traffic by ensuring that ATC capacity is utilized to the maximum extent possible, and that the traffic volume is compatible with the capacities declared by the appropriate ATS authority.

空域管理 airspace management

The most effective exploitation of the airspace in accordance with the requirements of the various airspace users.

空中交通服务 air traffic service

A generic term meaning variously, flight information service, alerting service, air traffic control service (area control service, approach control service or aerodrome control service).

空中交通服务报告室 air traffic services reporting office

A unit established for the purpose of receiving reports concerning air traffic services and flight plans submitted before departure.

航路 airway

A control area or part of a control area established in the form of a corridor equipped with

Unit One
Basic Operating Procedures

radio navigation aids.

告警服务 alerting service

A service provided to notify appropriate organizations regarding aircraft in need of search and rescue aid, and assist such organizations as required.

备降机场 alternate airport

An airport at which an aircraft may land if a landing at the intended airport becomes inadvisable.

高度 altitude

The vertical distance of an object measured from mean sea level.

进近管制单位 approach control unit

A unit established to provide air traffic control service to controlled flights arriving at, or departing from, one or more aerodromes.

进近管制服务 approach control service

ATC service for arriving or departing or transiting controlled flights.

进近顺序 approach sequence

The order in which two or more aircraft are cleared to approach to land at the aerodrome.

机坪 apron

Any area of an airport set aside for parking, maintenance, loading/unloading, etc. of aircraft; as opposed to runways.

机坪管理单位 apron management unit

A unit responsible for providing ground transportation service on the apron.

区域管制单位 area control office

A unit established to provide air traffic control service to controlled flights in control areas under its jurisdiction.

空中交通服务航路 ATS route

A specified route designed for channeling the flow of traffic as necessary for the provision of air traffic services.

自动相关监视 automatic dependent surveillance

A surveillance technique in which aircraft automatically provide, via a data link, data derived from on-board navigation and position fixing systems, including aircraft identification, four-dimensional position and additional data as appropriate.

通播 automatic terminal information service

The automatic provision of current, routine information to arriving and departing aircraft by means of continuous and repetitive broadcasts throughout the day or a specified portion of the day.

广播 broadcast

A transmission of information relating to air navigation that is not addressed to a specific

Lesson One

Terms, Commonly Used Abbreviations, Numbers, Call Signs, Phonetic Alphabets, Standard Words and Phrases

station or stations.

基线转弯 base turn

A turn executed by the aircraft during the initial approach between the end of the outboard track and the beginning of the intermediate or final approach track.

盲发 blind transmission

A transmission from one station to another station in circumstances where two-way communication cannot be established but where it is believed that the called station is able to receive the transmission.

盘旋进近 circling approach

An extension of an instrument approach procedure which provides for visual circling of the aerodrome prior to landing.

许可界限 clearance limit

The point to which an aircraft is granted an air traffic control clearance.

管制区 control area

A controlled airspace extending upwards from a specified limit above the earth.

管制地带 control zone

A controlled airspace extending upwards from the surface of the earth to a specified upper limit.

管制空域 controlled airspace

An airspace of defined dimensions within which air traffic control service is provided in accordance with the airspace classification.

巡航高度层 cruising level

A level maintained during a significant portion of a flight.

决断高度/决断高 decision altitude/height

A specified altitude or height in a 3D instrument approach operation at which a missed approach must be initiated if the required visual reference to continue the approach has not been established.

Note:(1) Decision altitude (DA) is referenced to mean sea level and decision height (DH) is referenced to the threshold elevation.

(2) The required visual reference means that section of the visual aids or of the approach area which should have been in view for sufficient time for the pilot to have made an assessment of the aircraft position and rate of change of position, in relation to the desired flight path. In Category Ⅲ operations with a decision height the required visual reference is that specified for the particular procedure and operation.

(3) For convenience where both expressions are used they may be written in the

form "decision altitude/ height" and abbreviated "DA/H".

偏航 deviation

The situation in which the planned track differs from the actual track.

遇险 distress

A situation wherein there is a reasonable certainty that an aircraft and its occupants are threatened by grave and imminent danger or require immediate assistance.

标高 elevation

The vertical distance of a point or level on, or affixed to, the surface of the earth measured from mean sea level.

续航能力 endurance

The time an aircraft can continue flying, without refueling.

预计到达时间 estimated time of arrival

The time at which the pilot estimates that the aircraft will be over a specific location.

预计进近时间 expected approach time

The time at which ATC expects that an arriving aircraft, following a delay, will leave the holding fix to complete its approach for a landing.

Note: The actual time of leaving the holding fix will depend upon the approach clearance.

最后进近 final approach

That part of an instrument approach procedure which commences at the specified final approach fix or point, or where such a fix or point is not specified,

(1) at the end of the last procedure turn, base turn or inbound turn of a racetrack procedure, if specified; or

(2) at the point of interception of the last track specified in the approach procedure; and ends at a point in the vicinity of an aerodrome from which:

a. a landing can be made; or

b. a missed approach procedure is initiated.

定位点 fix

A geographical position determined by visual reference to the surface, by reference to one or more radio NAVAIDs, by celestial plotting, or by another navigational device.

高度层 flight level

A surface of constant atmosphere pressure which is related to a specific pressure datum, 1013.2hPa.

飞行计划 flight plan

Specified information provided to air traffic services units, relative to an intended flight or portion of a flight of an aircraft.

航向 heading

The direction in which the longitudinal axis of an aircraft is pointed, expressed in

Lesson One
Terms, Commonly Used Abbreviations, Numbers, Call Signs, Phonetic Alphabets, Standard Words and Phrases

clockwise degrees from North. Usually magnetic heading (with respect to the magnetic north pole), sometimes true heading (with respect to the geographic north pole).

高 **height**

The vertical distance of a level, a point, or an object considered as a point, measurement from a specified datum.

等待 **holding**

In air traffic control, a predetermined maneuver which keeps aircraft within a specified airspace while awaiting further clearance from air traffic control.

等待定位点 **holding fix**

A specified fix identifiable to a pilot by NAVAIDs or visual reference to the ground used as a reference point in establishing and maintaining the position of an aircraft while holding.

等待点 **holding point**

An area where pilots wait for take-off clearance. It is an expression used in radiotelephony phraseology having the same meaning as Taxiway Holding Position or Runway Holding Position.

等待程序 **holding procedure**

A predetermined maneuver which keeps an aircraft within a specified airspace whilst awaiting further clearance.

识别 **identification**

The situation which exists when the position indication of a particular aircraft is seen on a situation display and positively identified.

仪表进近程序 **instrument approach procedure**

A series of predetermined maneuvers by reference to flight instruments with specified protection from obstacles from the initial approach fix, or when applicable, from the beginning of a defined arrival route to a point from which the landing can be completed and thereafter, if a landing is not completed, to a position at which holding or en-route obstacle clearance criteria apply.

仪表飞行规则 **instrument flight rules**

A set of rules governing the conduct of flight under instrument meteorological conditions.

仪表气象条件 **instrument meteorological conditions**

Meteorological conditions expressed in terms of visibility, horizontal and vertical distance from cloud, less than the minima specified for visual meteorological conditions.

仪表飞行 **IFR flight**

A flight conducted in accordance with IFR.

非全跑道离场 **intersection departure**

A departure from any runway intersection except the end of the runway.

着陆区 **landing area**

That part of a movement area intended for the landing or take-off of aircraft.

高度 level

A generic term relating to the vertical position of an aircraft in flight and meaning variously, height, altitude or flight level.

机动区 maneuvering area

The part of an aerodrome to be used for the take-off, landing and taxiing of aircraft, excluding aprons.

微波进近 microwave approach

Approach made by aircraft under the guidance of microwave landing system.

最低下降高度/高 minimum descent altitude/height

An altitude/height in a non-precision or circling approach below which descent may not be made without visual reference.

复飞点 missed approach point

The point in an instrument approach procedure at or before which the prescribed missed approach procedure must be initiated in order to ensure that the minimum obstacle clearance is not infringed.

复飞程序 missed approach procedure

The procedure to be followed if the approach cannot be continued.

活动区 movement area

The part of an aerodrome to be used for the take-off, landing and taxiing of aircraft, including aprons.

非精密进近程序 non-precision approach procedure

A standard instrument approach procedure in which no electronic glide slope is provided, such as VOR and NDB.

精密进近程序 precision approach procedure

A standard instrument approach procedure in which an electronic glide slope is provided, such as ILS and PAR.

程序转弯 procedure turn

A maneuver in which a turn is made away from a designated track followed by a turn in the opposite direction to permit the aircraft to intercept and proceed along the reciprocal of the designated track.

Note: (1) Procedure turns are designated "left" or "right" according to the direction of the initial turn.

(2) Procedure turns may be designated as being made either in level flight or while descending, according to the circumstances of each individual approach procedure.

(3) There are two kinds of procedure turn: 45°/180° and 80°/260°.

直角航线程序 racetrack procedure

A procedure designed to enable the aircraft to reduce altitude during the initial approach

Lesson One
Terms, Commonly Used Abbreviations, Numbers, Call Signs, Phonetic Alphabets, Standard Words and Phrases

segment and/or establish the aircraft inbound when the entry into a reversal procedure is not practical.

雷达看到 radar contact

The situation which exists when the radar blip or radar position symbol of a particular aircraft is seen and identified on a radar display, and radar services are being provided.

雷达识别 radar identification

The situation which exists when the radar position of a particular aircraft is seen and identified on a situation display.

雷达引导 radar vectoring

Provision of navigational guidance to aircraft in the form of specific headings, based on the use of radar.

救援协调中心 rescue co-ordination center

A search and rescue (SAR) facility equipped and manned to coordinate and control SAR operations in an area designated by the SAR plan.

限制区 restricted area

An airspace of defined dimensions, above the land areas or territorial waters of a State, within which the flight of aircraft is restricted in accordance with certain specified conditions.

反向程序 reverse procedure

A procedure designed to enable aircraft to reverse direction during the initial approach segment of an instrument approach procedure.

Note: Reverse procedure includes procedure turns or base turns.

跑道 runway

A defined rectangular area on a land aerodrome prepared for the landing and take-off of aircraft.

道面活动 runway movement

Any movement of aircraft on available runway(s).

跑道视程 runway visual range

The range over which the pilot of an aircraft on the centerline of a runway can expect to see the runway surface markings, or the lights delineating the runway or identifying its centerline.

直线进近 straight-in approach

An instrument approach wherein final approach is commenced without first having executed a procedure turn. (Not necessarily completed with a straight-in landing)

终端管制区 terminal control area

A control area normally established at the confluence of ATS routes in the vicinity of one or more major aerodromes.

入口 threshold
The beginning of that portion of the runway usable for landing.

接地点 touchdown
The point where the nominal glide path intercepts the runway.

空中防撞系统 traffic alert and collision avoidance system
参见 airborne collision avoidance system.

过渡高度 transition altitude
The altitude at or below which the vertical position of an aircraft is controlled by reference to altitudes.

过渡高度层 transition level
The lowest flight level available for use above the transition height (altitude).

过渡高 transition height
The height at or below which the vertical position of an aircraft is controlled by reference to heights.

紧急 urgency
An urgent condition, one of being concerned about safety, and requiring timely but not immediate assistance.

能见度 visibility
The ability, as determined by atmospheric conditions and expressed in units of distance, to see and identify prominent unlighted objects by day and prominent lighted objects by night.

目视进近 visual approach
An approach conducted under IFR that authorizes the pilot to proceed visually and clear of clouds to the airport.

目视气象条件 visual meteorological conditions
Meteorological conditions expressed in terms of visibility, horizontal and vertical distance from cloud, equal to or better than specified minima.

目视间隔 visual separation
To maintain the orderly flow of air traffic and avoid the collision between aircraft, the controller sees the aircraft involved, or the pilot sees the other aircraft involved in order to maintain safe separation between aircraft.

尾流 wake turbulence
Phenomena resulting from the passage of an aircraft through the atmosphere. The term includes vortices, thrust stream turbulence, jet blast, jet wash, propeller wash, and rotor wash both on the ground and in the air.

Lesson One
Terms, Commonly Used Abbreviations, Numbers, Call Signs, Phonetic Alphabets, Standard Words and Phrases

2. Commonly Used Abbreviations(常用缩略语)

ACC	Area Control Center
ADF	Automatic Direction-finding Equipment
ADS	Automatic Dependent Surveillance
AFIS	Aerodrome Flight Information Service
AGL	Above Ground Level
AIC	Aeronautical Information Circular
AIP	Aeronautical Information Publication
AIRAC	Aeronautical Information Regulation and Control
AIS	Aeronautical Information Service
AMSL	Above Mean Sea Level
AOM	Airport Operation Minima
AOR	Area of Responsibility
APU	Auxiliary Power Unit
ATC	Air Traffic Control
ATD	Actual Time of Departure
ATFM	Air Traffic Flow Management
ATIS	Automatic Terminal Information Service
ATS	Air Traffic Service
ATZ	Aerodrome Traffic Zone
CAVOK	Ceiling and Visibility OK, i.e. visibility, cloud and present weather better than prescribed values or condition
CIP	Commercially Important Person
CPDLC	Controller-Pilot Data Link Communications
CRP	Compulsory Reporting Point
CNS	Communications, Navigation, Surveillance
CTR	Control Zone
DME	Distance Measuring Equipment
EAT	Expected Approach Time
EET	Estimated Elapsed Time
ETA	Estimated Time of Arrival
ETD	Estimated Time of Departure
ETO	Estimated Time Over
FAA	Federal Aviation Administration
FIC	Flight Information Center

13

Unit One
Basic Operating Procedures

FIR	Flight Information Region
FIS	Flight Information Service
GNSS	Global Navigation Satellite System
GPS	Global Positioning System
HF	High Frequency
H24	Continuous Day and Night Service
IFR	Instrument Flight Rules
ILS	Instrument Landing System
IMC	Instrument Meteorological Condition
INFO	Information
INS	Inertial Navigation System
LORAN	Long Range Air Navigation System
MET	Meteorological or Meteorology
MLS	Microwave Landing System
MNPS	Minimum Navigation Performance Specifications
NDB	Non-directional radio Beacon
NOZ	Normal Operating Zone
NTZ	No-Transgression Zone
NIL	None or I have nothing to send you
NOTAM	Notice to Airman, i. e. a notice containing information concerning the establishment, condition or change in any aeronautical facility, service procedure or hazard, the timely knowledge of which is essential to personnel concerned with flight operations.
PAOAS	Parallel Approach Obstacle Assessment Surfaces
PBN	Performance-Based Navigation
QFE	Atmospheric pressure at aerodrome elevation/at runway threshold
QNH	Altimeter sub-scale setting to obtain elevation when on the ground
RCC	Rescue Co-ordination Center
RNAV	Area Navigation
RNP	Required Navigation Performance
RVR	Runway Visual Range
RVSM	Reduced Vertical Separation Minimum
SELCAL	A system which permits the selective calling of individual aircraft over radiotelephone channels linking a ground station with the aircraft.
SID	Standard Instrument Departure
SIGMET	Information concerning en-route weather phenomena which may affect safety of aircraft operations

Lesson One

Terms,Commonly Used Abbreviations,Numbers,Call Signs,Phonetic Alphabets,Standard Words and Phrases

SNOWTAM	A special series NOTAM notifying the presence or removal of hazardous conditions due to snow, ice, slush or standing water associated with snow, slush and ice on the movement area, by means of a special format.
SPECI	Special Aviation Report
SSR	Secondary Surveillance Radar
SST	Supersonic Transport
STAR	Standard Terminal Arrival Route
TCAS 或 ACAS	Traffic Alert and Collision Avoidance System/Airborne Collision Avoidance System
TAF	Aerodrome Forecast
TMA	Terminal Control Area
UHF	Ultra-High Frequency
UIR	Upper Flight Information Region
UTA	Upper Control Area
UTC	Coordinated Universal Time
VASIS	Visual Approach Slope Indicator System
VFR	Visual Flight Rules
VHF	Very High Frequency
VIP	Very Important Person
VMC	Visual Meteorological Conditions
VOLMET	Meteorological information for aircraft in flight
VOR	VHF Omni-directional Radio Range

3. Numbers(数字)

1) General

Number	Pronunciation	汉语读法
0	ZE-RO	洞
1	WUN	幺
2	TOO	两
3	TREE	三
4	FOW-er	四
5	FIFE	五
6	SIX	六
7	SEV-en	拐

15

Unit One
Basic Operating Procedures

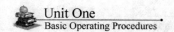

8	AIT	八
9	NIN-er	九
Decimal	DAY-SEE-MAL	点
Hundred	HUN-dred	百
Thousand	TOU-SAND	千

注：英文大写部分应重读。

When transmitting messages containing aircraft call signs, altimeter settings, flight levels, headings, wind speeds/directions, pressure settings, transponder codes and frequencies, ATC/pilots shall transmit each digit separately.

All numbers used in the transmission of altitude, height, cloud height, visibility and runway visual range information which contain whole hundreds and whole thousands shall be transmitted by pronouncing each digit in the number of hundreds or thousands followed by the word HUNDRED or TOUSAND as appropriate. Combinations of thousands and whole hundreds shall be transmitted by pronouncing each digit in the number of thousands followed by the word THOUSAND and the number of hundreds followed by the word HUNDRED.

10	WUN ZE-RO(幺洞)
79	SEV-en NIN-er(拐九)
300	TREE HUN-dred(三百)
13700	WUN TREE TOU-SAND SEV-en HUN-dred(一万三千七百)
341	TREE FOW-er WUN(三四幺)
68539	SIX AIT FIFE TREE NIN-er(六八五三九)
BAW 246	SPEEDBIRD TOO FOW-er SIX(英航两四六)
150 degrees	WUN FIFE ZE-RO DEGREES(一百五十度或幺五洞)
180 knots	WUN AIT ZE-RO KNOTS(速度幺八洞)
12m/s	WUN TOO METERS PER SECOND(十二米秒)
118.65	WUN WUN AIT DAY-SEE-MAL SIX FIFE(幺幺八点六五)
123.7	WUN TOO TREE DAY-SEE-MAL SEV-en(幺两三点拐)
QNH 1013	QNH WUN ZE-RO WUN TREE(修正海压幺洞幺三)
QFE 998	QFE NIN-er NIN-er AIT(场压九九八)
heading 160	HEADING WUN SIX ZE-RO(航向幺六洞)
6300m	SIX TOU-SAND TREE HUN-dred METERS(六千三)
7200m	SEV-en TOU-SAND TOO HUN-dred METERS(拐两)
11000m	WUN WUN TOU-SAND METERS(幺幺洞)
14300m	WUN FOW-er TOU-SAND TREE HUN-dred METERS(幺四三)
3900ft	TREE TOU-SAND NIN-er HUN-dred FEET(三千九百英尺)
27000ft	FLIGHT LEVEL TOO SEV-en ZE-RO(高度层两拐洞)

Lesson One
Terms,Commonly Used Abbreviations,Numbers,Call Signs,Phonetic Alphabets,Standard Words and Phrases

2) Transmission of time

When transmitting time, only the minutes of the hour are normally required. However, the hour should be included if there is any possibility of confusion. Time checks shall be given to the nearest minute. Coordinated Universal Time (UTC) is to be used at all times, unless specified. 2400 hours designates midnight, the end of the day, and 0000 hours the beginning of the day.

0735	TREE FIFE or ZE-RO SEV-en TREE FIFE
2321	TOO WUN or TOO TREE TOO WUN
1200	WUN TOO ZE-RO ZE-RO
1930	TREE ZE-RO or WUN NIN-er TREE ZE-RO

注1：通报时间一般默认为协调世界时（UTC），如通报时间为北京时（Beijing time），应特殊说明。

注2：当机组觉得必要时可向管制单位申请校时（request time check），此时将实际时间精确到最接近的半分钟。

3) Transmission of airport elevation

汉语读法为：标高+数字+单位

英语读法为：elevation + number + measurement unit, e. g.

XY 机场标高 138 米：XY AIRPORT ELEVATION WUN TREE AIT METERS

北京首都国际机场标高 35 米：BEIJING CAPITAL INTERNATIONAL AIRPORT ELEVATION TREE FIFE METERS

4. Call signs（呼号）

1) Call signs for Aeronautical Stations

管制单位或服务	后缀汉语简呼	后缀英语简呼
区域管制中心 （Area control center）	区域	CONTROL
进近管制 （Approach control）	进近	APPROACH
进场雷达管制 （Approach control radar arrival）	进场	ARRIVAL
离场雷达管制 （Approach control radar departure）	离场	DEPARTURE
机场管制 （Aerodrome control）	塔台	TOWER

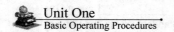
Unit One Basic Operating Procedures

地面活动管制 (Surface movement control)	地面	GROUND
放行许可发布 (Clearance delivery)	放行	DELIVERY
飞行服务/航空情报服务 (Flight information service)	飞服	INFORMATION
精密进近雷达 (Precision approach radar)	精密	PRECISION
机坪管制/管理服务 (Apron control/management service)	机坪	APRON
公司签派 (Company dispatch)	签派	DISPATCH
航空电台 (Aeronautical station)	电台	RADIO
五边监控席 (Final approach radar monitoring)	五边	FINAL

2) Call signs for aircraft

The call signs of aircraft are divided into five categories:

(1) Five-letter aircraft registration code, e. g. GBDFT

(2) Operator's radio code+five-letter aircraft registration code, e. g. BAW ABCDE

(3) Aircraft type+aircraft registration code, e. g. Piper GBSZT

(4) Operator's radio code+flight number, e. g. CSN 306

(5) National registration code+numbers, e. g. N310295

After satisfactory communication has been established, and provided that no confusion is likely to occur, the ground station may abbreviate call signs (see table below). A pilot may only abbreviate the call sign of his aircraft if it has first been abbreviated by the aeronautical station.

Full Call sign	FBACD	SABENA GCDAB	Piper GBSZT	ACA 165	N416732
Abbreviated Call sign	FCD	SABENA AB	Piper ZT	ACA 165	N732

Terms,Commonly Used Abbreviations,Numbers,Call Signs,Phonetic Alphabets,Standard Words and Phrases

5. Phonetic Alphabet(字母)

字母 LETTER	单词 WORD	发音 PRONUNCIATION
A	Alpha	AL FAH
B	Bravo	BRAH VOH
C	Charlie	CHAR LEE
D	Delta	DELL TAH
E	Echo	ECK OH
F	Foxtrot	FOKS TROT
G	Golf	GOLF
H	Hotel	HOH TELL
I	India	IN DEE AH
J	Juliett	JEW LEE ETT
K	Kilo	KEY LOH
L	Lima	LEE MAH
M	Mike	MIKE
N	November	NO VEM BER
O	Oscar	OSS CAH
P	Papa	PAH PAH
Q	Quebec	KEH BECK
R	Romeo	ROW ME OH
S	Sierra	SEE AIR RAH
T	Tango	TANG GO
U	Uniform	YOU NEE FORM

V	Victor	VIK TAH
W	Whiskey	WISS KEY
X	X-ray	ECKS RAY
Y	Yankee	YANG KEY
Z	Zulu	ZOO LOO

注：下画线部分应重读。

6. Standard Words and Phrases（标准词和词组）

ACKNOWLEDGE 请认收

Let me know that you have received and understood this message.

CTL：AAL 981, acknowledge the following ATC clearance.

AFFIRM 是的

Yes.

CTL：JAL 065, are you ready for taxi?

PIL：Affirm, JAL 065.

APPROVED 同意

Permission for proposed action granted.

PIL：Tower, CES 2837, request start-up and push-back.

CTL：CES 2837, start-up and push-back approved.

BREAK 还有

Indicates the separation between messages.

CTL：BAW 715, climb to 6300m. Break, there is a previous pilot report he encountered moderate turbulence between 4500m and 5100m.

BREAK BREAK

Indicates the separation between messages transmitted to different aircraft in a busy environment.

CTL：JAL 512, descend to 3000m immediately and report number of souls on board. Break Break, all stations, stop transmitting, aircraft executing emergency descent.

注：中文通话仍然使用"BREAK BREAK"。

CANCEL 取消

Annul the previously transmitted clearance.

CTL：CCA 1509, Dalian Tower, hold position, cancel take-off, I say again, cancel take-

Lesson One

Terms, Commonly Used Abbreviations, Numbers, Call Signs, Phonetic Alphabets, Standard Words and Phrases

off, unknown traffic taxiing onto the runway.

CHECK 检查

Examine a system or procedure. (Not to be used in any other context. No answer is normally expected.)

PIL: Beijing Tower, KAL 036, request time check.

CTL: KAL 036, Beijing Tower, time 1206 and a half.

CLEARED 可以

Authorized to proceed under the conditions specified.

CTL: SIA 901, wind 210, 8m/s, runway 24, cleared to land.

CONFIRM 证实

Have I correctly received the following…? or

Did you correctly receive this message?

(1) CTL: SAS 817, confirm proceeding to parking stand Z5.

(2) CTL: CSZ 8253, confirm leaving 8400m for 7200m.

CONTACT 联系

Establish communications with… (your details have been passed).

CTL: CSC 7682, contact Guangzhou Tower on 119.3.

CORRECT 正确

That is correct.

CTL: DAL 017, read back correct.

CORRECTION 更正

An error has been made in this transmission (or message indicated). The correct version is…

PIL: Beijing Control, DLH 837, VYK 34, correction, VYK 38, 6300 meters.

DISREGARD 作废

Consider that transmission as not sent.

CTL: UAL 601, Hong Kong Delivery, disregard previous ATC clearance.

HOW DO YOU READ 你听我几个

What is the readability of my transmission?

PIL: Harbin Tower, KAL 736, radio check on 118.3, how do you read?

I SAY AGAIN 我重复一遍

I repeat for clarity or emphasis.

PIL: Beijing Approach, CSN 3509, 15km north of VYK, passenger suspected heart attack, I say again, passenger suspected heart attack, request priority landing and medical service on arrival.

MONITOR 守听

Listen out on (frequency).

21

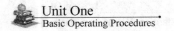
Unit One
Basic Operating Procedures

CTL: All stations, monitor Ground on 123.45, MAYDAY.

NEGATIVE 错误,不同意或没有

No; or Permission not granted; or That is not correct.

CTL: CHH 7629, Nanjing Approach, climb straight ahead until 300 meters before turning left.

PIL: Climbing straight ahead to 600 meters and then turn left, CHH 7629.

CTL: CHH 7629, negative, climb straight ahead to 300 meters before turning left.

OUT 完毕

This exchange of transmissions is ended and no response is expected.

(The word "OUT" is not normally used in VHF communications.)

OVER 请回答

My transmission is ended and I expect a response from you.

(The word "OVER" is not normally used in VHF communications.)

READ BACK 复诵

Repeat all, or the specified part, of this message back to me exactly as received.

CTL: UAL 765, read back correct.

RECLEARED 重新许可

A change has been made to your last clearance and this new clearance supersedes your previous clearance or part thereof.

CTL: JAL 589, Guangzhou Control, recleared for 10400m, report reaching.

REPORT 报告

Pass me the following information.

CTL: KLM 109, report runway in sight.

REQUEST 申请

I should like to know… or I wish to obtain…

PIL: Pudong Tower, VIR 714, holding at holding point runway 35, request line-up.

ROGER 收到

I have received all your last transmission.

Note: Under no circumstances to be used in reply to a question requiring a direct answer in the affirmative (AFFIRM) or negative (NEGATIVE).

PIL: Zhuhai Approach, ANA 491, bird ingestion on port engine during climbing, request return for landing.

CTL: ANA 491, roger.

SAY AGAIN 重复一遍

Repeat all, or the following part of your last transmission.

CTL: QFA 976, Shanghai Control, say again your troubles and report number of persons on board.

Lesson One

Terms, Commonly Used Abbreviations, Numbers, Call Signs, Phonetic Alphabets, Standard Words and Phrases

SPEAK SLOWER 讲慢点

Reduce your rate of speech.

CTL: BAW G-CDAB, say again your intentions and speak slower.

STANDBY 稍等

Wait and I will call you.

PIL: Ground, N342178, request push-back.

CTL: N178, standby, no tug is available at the moment.

UNABLE 不能

I cannot comply with your request, instruction, or clearance.

CTL: CCA 101, Beijing Control, climb and maintain 11000m.

PIL: Unable due weight, CCA 101.

注:"UNABLE"或"UNABLE TO COMPLY"后面通常要跟原因。

VERIFY 核实

Request confirmation of information.

CTL: CAL 106, Wuhan Ground, verify your MTOW with your company.

WILCO 照办

I understand your message and will comply with it.

(abbreviation for "will comply")

CTL: CXA 8167, Qingdao Tower, expedite vacating runway, traffic on short final.

PIL: CXA 8167, wilco, expediting.

Note: The opposite for "Wilco" is "Unable" or "Unable to comply".

WORDS TWICE 讲两遍

(1) As a request: Communication is difficult. Please send every word twice.

PIL: Shanghai Control, CCA 1538, I read you 3, words twice.

(2) As Information: Since communication is difficult, every word in this message will be sent twice.

CTL: CES 2163, Shenzhen Tower, you are cut in and out, words twice.

Notes

1. The pronunciations of numbers used in aviation may sound strange to the non-aviation person, but they are used for a very important reason. Transmissions on radios are not always clear due to many factors such as interference, poor equipment, cockpit noise, and the regional influences of the speakers. In the infancy of radiotelephony, certain numbers were discovered to be more difficult to distinguish than others. For example, in an environment of noise and interference, the words "two" and "three" can sound very similar. By using "tree" instead of "three," the r sound is clearer and makes the distinction easier.

Similarly, the r sound is much stronger in the pronunciation "fower" than in "four." The

Unit One
Basic Operating Procedures

words "five" and "nine" are sometimes confused, especially on poor quality radios; so the pronunciations "fife" and "niner" are used to make the distinction obvious. The word "thousand" can sound like "how," so "tousand" is used in order to make the t sound stronger. The reason that most multi-digit numbers are spoken one digit at a time is because of the tendency of "eleven" to sound like "seven" and of many two-digit numbers to sound similar, as in "fifteen" and "fifty," "seventeen" and "seventy," and so on.

2. In radiotelephony communications, the alphabets should all be pronounced according to ICAO requirements, namely the phonetic alphabets listed above. However, exceptions are as follows, if no misunderstanding is incurred:

(1) Some routine alphabet combinations are still pronounced as usual, e.g.: NDB, ILS, VOR, GPS;

(2) Airlines call signs are pronounced as registered, e.g. CES is pronounced as China Eastern, BAW as Speedbird, and KLM as KLM;

(3) Aircraft type is pronounced as its registration type, e.g. B747 is pronounced as Boeing 747, EMB 145 as EMB 145, MD-11 as MD eleven.

Exercises

I. Read the following aloud.

1. QWERI MKJHX IOPGF LAWPX OEJDB VMBZA GFVQK
   FUYIMNCA SWEQZSETLV PODHXBKGRIJLNZ
2. 2100m 5400m 9500m 10700m 11300m 12200m
   22600ft 30100ft 37100ft 46900ft 10800ft
3. 123.5 116.7 113.4 120.8 118.6 121.75 119.37 124.35
4. 0130 1428 2236 0749 1513 2300 1602 1354

II. Give the abbreviated forms of the following call signs.

DIOMH BOWPD AFR 901 CDG 573 Shorts NKIAC Cessna BZXIG
BAW LIGPC ABB GABCD N564582 Q259741

III. Dictation: Write down what you hear on the recording.

1.
2.
3.
4.
5.

Lesson 1-Exercise 3

Lesson Two
Establishing Communication and Transfer of Communication

Model Exchanges

Lesson 2

Establishing communication

A.
PIL: Pudong Ground, CSN 3392, request departure information.
CTL: CSN 3392, Pudong Ground, departure runway 17, wind 150 degrees, 4m/s, QNH 1008, temperature minus 2 degrees centigrade, dew point minus 5 degrees centigrade, RVR 550 meters.
PIL: Runway 17, QNH 1008, will call for start-up, CSN 3392.

B.
PIL: Beijing Approach, CES 2987, 6600m, descending to 4200m.
CTL: CES 2987, Beijing Approach, roger.

C.
CTL: Station calling Wuhan Tower, say again your call sign.
PIL: Wuhan Tower, CCA 1509, ready for taxi.

Transfer of communication

D.
CTL: CCA 1537, contact Shanghai Approach on 119.75.
PIL: 119.75, CCA 1537.

Unit One
Basic Operating Procedures

E.
PIL: Shanghai Approach, CHH 8531, we have established contact with Hefei Control, good day.

F.
CTL: UAL 031, standby for Hongqiao Tower on 118.1.
PIL: 118.1, UAL 031.

G.
CTL: JAL 781, if no contact, monitor Ground on 121.3.

Dialogue

A.
PIL: Beijing Delivery, …9217.
CTL: Station calling Beijing Delivery, say again your call sign.
PIL: Beijing Delivery, CSC 9217.
CTL: CSC 9217, standby.
PIL: Standing by, CSC 9217.
CTL: CSC 9217, pass your message.
PIL: CSC 9217, destination Chengdu, stand 7, request ATC clearance.

B.
CTL: SIA 931, runway 21, cleared for take-off.
PIL: Runway 21, cleared for take-off, SIA 931.
CTL: SIA 931, contact Echo Departure on 119.1.
PIL: 119.1, SIA 931.
PIL: Echo Departure, SIA 931.
CTL: SIA 931, continue climb to 3000 meters.
PIL: Climbing to 3000 meters, SIA 931.
PIL: MAYDAY MAYDAY MAYDAY, Echo Departure, SIA 931, we have to return to land, smoke coming out from No. 2 engine.
CTL: SIA 931, descend to 1200 meters, cleared for ILS approach runway 21. Break Break, all stations, stop transmitting, MAYDAY.

New Words and Phrases

information *n.* 情报,消息,资料,通播
departure information 离场条件
dew point 露点
transfer *v.* 移交
establish *v.* 建立(盲降、联系等)

Lesson Two
Establishing Communication and Transfer of Communication

RVR（Runway Visual Range）跑道能见视程，读作 R-V-R
ATIS（Automatic Terminal Information Service）航站自动情报服务，简称"通播"
ILS（Instrument Landing System）仪表着陆系统，俗称"盲降"
transmit *v.* 发送，传递
stop transmitting 无线电静默
delivery *n.* 放行
destination *n.* 目的地，终点
ATC clearance 空中交通管制放行许可
MAYDAY *n.* 遇险

Notes

1. 空中交通无线电通话用语应用于空中交通服务单位与航空器之间的话音联络。它有自己特殊的发音规则，语言简洁、严谨，经过严格的缩减程序，通常为祈使句。

2. 通话结构。

（1）首次联系时，空中交通管制员采用的通话结构应为：
对方完整呼号+己方呼号+通话内容

（2）首次通话以后的各次通话，空中交通管制员可以采用下列通话结构：
对方呼号+通话内容

（3）航空器驾驶员采用的通话结构应为：
对方呼号+己方完整呼号+通话内容

（4）空中交通管制员确认航空器驾驶员复诵的内容正确时，通话结构应为：
对方呼号+"（复诵）正确"

注：航空器驾驶员应以完整呼号终止复诵。

3. 当地面电台或者某一航空器想对周围的航空器广播信息或情报时，可在信息或者情报前加上"全体注意（All stations）"。

4. 如果被呼叫单位不能确定谁呼叫自己，被呼叫单位可以要求对方重复呼号直至建立联系。

5. 当航空器需要从一个无线电频率转换到另一个频率时，管制单位应通知航空器转换频率。如果管制单位未通知，驾驶员应在转换频率之前提醒管制员。

6. 当管制单位需要与航空器进一步通话时，可指示航空器在某频率上守听。

Exercises

I. Oral practice.

Use the call signs in column A to make initial contact with the aeronautical stations in column B.

    A.   B.
CCA 1573   Pudong Tower

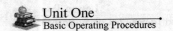
Unit One
Basic Operating Procedures

  Boeing N376012   Beijing Control
  F-GCIK   Guanghan Tower
  JAL GCABD   Nanjing Approach
  CSN 3129   Zhuhai Approach

II. Fill in the blanks.

1. PIL: Hongqiao Tower, CES 2842.

CTL: CES 2842, Hongqiao Tower, _____ .

2. CTL: _____ Beijing Approach, _____ your call sign.

3. CTL: CDG 9145, _____ Tower on 117.2.

PIL: _____ , CDG 9145.

4. CTL: CSN 3507, climb to 6600 meters, report passing 3600 meters.

PIL: Roger, climb to 6600 meters, call you 3300 meters, _____ , report 3600 meters.

5. CTL: All stations, _____ Ground on 118.75, MAYDAY.

6. The pilot wanted to know the readability of his transmission so he asked _____ ?

7. It looked like the aeroplane was going to park at the wrong gate, so I asked the pilot: S-DE, _____ you are proceeding to stand 7.

8. I sent the information to the wrong aircraft, so I quickly told the pilot to _____ my last message.

9. When I cannot answer a pilot at once, I tell him to _____ .

10. The pilot requested to make a right turn after take-off. I had no objections, so I said: Right turn out is _____ .

III. Dictation: Write down what you hear on the recording.

1.

2.

3.

4.

5.

6.

Lesson 2-
Exercise 3

Unit Two

Preflight to Line-up

Lesson Three

ATIS, Arrival/Departure Information, Essential Aerodrome Information, VOLMET

Model Exchanges

Lesson 3

ATIS

A.

Hong Kong International Airport information D at time 0030 UTC, landing runway 07L, runway surface wet, wind 080 degrees 14 knots, gusting 23 knots, visibility 4000 meters, showers, clouds few ceiling 700 feet, scattered ceiling 1800 feet, broken ceiling 5000 feet, temperature 26 degrees centigrade, dew point 17 degrees centigrade, QNH 1004 hectopascals. Tempo visibility 3000 meters, light showers. Advise on initial contact you have information D.

B.

Nanjing Lukou International Airport, information H, 1350 UTC, ILS approach runway 06, runway surface wet, braking action medium, wind calm, CAVOK, temperature plus 3 degrees centigrade, dew point minus 2 degrees centigrade, QNH 1016. Caution, work in progress near taxiway E, and traffic is crowded near apron. On initial contact, advise you have received information H.

Arrival/Departure information

C.

PIL: Beijing Ground, CES 2398, request arrival information.

CTL: CES 2398, runway in use 36R, wind 350° 8m/s, gusting 13m/s, temperature 14 degrees centigrade, dew point 10 degrees centigrade, visibility 6km, runway is wet, braking action good, QNH 1023.

31

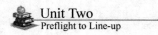
Unit Two
Preflight to Line-up

D.

PIL: Foxtrot Ground, UIA 358, request departure information.

CTL: UIA 358, departure runway 15, expect 3V departure, transition altitude 3000m, wind 170 degrees at 5kts, temperature 3 degrees centigrade, dew point minus 6 degrees centigrade, QNH 1018, caution flock of birds reported north of the field.

Essential aerodrome information

E.

CTL: ANA 010, Shenyang Ground, taxi with caution, snowdrifts up to 25cm adjacent to the intersection of taxiway S and T.

F.

CTL: CXA 9324, there is a broken-down vehicle parked near the jetway, caution the mechanics and equipment.

G.

CTL: CCA 1537, threshold of runway 18L displaced 180 meters due broken pavement.

H.

CTL: CQH 9166, braking action reported by previous landing A320 at 1835 medium to good.

I.

CTL: Runway surface condition, runway 24, available width 32m, covered with thin patches of ice, braking action poor, snow up to 30cm along edges.

J.

CTL: CSN 3178, taxi to runway 18L via taxiway E and M, caution, taxiway M centerline lighting unserviceable.

K.

CTL: CES 2157, ILS category II is unserviceable due glide slope antenna failure, land with caution.

VOLMETS

L.

This is Shanghai Hongqiao International Airport Volmet. 1500 UTC, wind 070, 4m/s, visibility 5000m, snow showers, clouds 2 octas ceiling 450m, temperature -1 degrees centigrade, dew point -5 degrees centigrade, QNH 1025, NO SIG, runway 18L damp up to 100%, braking action good, runway 18R wet up to 100%, braking action good.

M.

Beijing Met report, Beijing Capital International Airport 1530 Zulu Time, wind 280 degrees, 9m/s, visibility 1100m, RVR runway 36R 900m, drizzle, broken, ceiling 200m,

32

Lesson Three
ATIS,Arrival/Departure Information,Essential Aerodrome Information,VOLMET

temperature 8 degrees centigrade, dew point 6 degrees centigrade. Gradu, visibility 5000m, scattered, ceiling 400m.

New Words and Phrases

arrival *n*. 进场

departure *n*. 离场

knot *n*. 节,海里每小时

shower *n*. 阵雨

sky clear 碧空

few *adj*. 少云

scattered *adj*. 疏云

broken *adj*. 多云

overcast *adj*. 阴天

CAVOK(Ceiling and Visibility OK)天气晴好

altitude *n*. 高度(相对于修正海压)

tempo *adj*. 临时地(短时间内的天气状况)

gradu *adj*. 逐渐地

braking action 刹车效应

adjacent *adj*. 邻近的

intersection *n*. 道口

jetway *n*. 廊桥

mechanic *n*. 机械师,维修人员

alert *n*. 告警,提醒

threshold *n*. (跑道)入口

displaced *adj*. 内移的

threshold displaced 跑道入口内移(由跑道入口破损或污染等原因造成)

pavement *n*. (跑道或滑行道等)道面

braking action 刹车效应(一般有好、中好、中、中差和差五个等级)

standing water 积水

unserviceable *adj*. 失效,故障,不能提供服务

Notes

1. ATIS(Automatic Terminal Information Service)航站自动情报服务,简称通播。飞行量在年起降超过36000架次的机场,为了减轻空中交通管制甚高频通信波道的通信负荷,应当设立航站自动情报服务系统,为进、离场航空器提供服务。通播通常在一个单独的无线电频率上进行广播,包括主要的与飞行相关的信息,如天气、可用跑道、气压及高度表拨正值等信息。飞行员通常在和管制员等单位建立联系前收听通播,了解相关情况,以减少

33

管制员的工作量及避免频道拥挤。正常情况下通播每小时更新一次,天气变化迅速时也可以随时更新,依次以字母代码 A,B,C,…,Z 表示,按照 ICAO 公布的标准字母解释法判读。通播可分为进场通播、离场通播和进离场通播三种。

2. 通播的内容主要包括:
(1) 机场名称 airport name
(2) 代码 information code
(3) 观测时间 time
(4) 预计进近类别 estimated approach type
(5) 使用跑道 runway in use
(6) 重要的跑道道面情况,刹车效应 runway condition/braking action
(7) 延迟等待 delay
(8) 过渡高度层 transition level
(9) 其他必要的运行通报 operational information
(10) 地面风向量 wind direction (degrees) and strength (knots)
(11) 能见度,跑道视程 visibility, RVR
(12) 现行天气 present weather
(13) 云盖度,云底高 cloud cover in octas, height of clouds in feet or meters
(14) 大气温度 temperature
(15) 露点 dew point
(16) 高度表拨正值 altimeter
(17) 天气趋势 trend ('no sig' or expected weather change)
(18) 特殊信息 extra information

3. 机场情报(essential aerodrome information)是确保航空器在地面运行正常安全的必要信息,主要包含地面状况及相关设施的运行状况。主要分为:
(1) 活动区或者附近的施工和维修保养工作;
(2) 跑道滑行道的道面和附近状况,包括破损、污染物等;
(3) 临时性的障碍物或者危险状况;
(4) 灯光系统的非正常情况。

4. 跑道上污染物的描述。
damp,湿的
wet,潮的
water patches,块状积水
flooded,淹没
standing water,积水
pools of water,积水
water puddles,水洼
snowdrift,雪堆

Lesson Three
ATIS, Arrival/Departure Information, Essential Aerodrome Information, VOLMET

frozen ruts and ridges, 冻辙和冰脊

slush, 雪水

loose snow, 软雪

compacted snow, 压实的雪

dry snow, 干雪

firm snow, 实雪

wet snow, 潮雪

5. VOLMET 的内容主要包括：

（1）机场名称 airport name

（2）风向量 wind data（direction and strength）

（3）能见度 visibility（in meters or kilometers）

（4）当前天气 present weather（rain, mist, snow, drizzle, etc.）

（5）云盖度 cloud cover（in octas）

（6）云底高度 ceiling

（7）温度 temperature

（8）露点 dew point

（9）变化趋势 trend（no sig）

（10）逐渐变化趋势 gradu（plus expected change）

（11）临时变化情况 tempo（plus possible temporary conditions）

6. altitude——高度（修正海平面气压 QNH）；height——高（场面气压 QFE）；level——高度层（标准气压 QNE）。

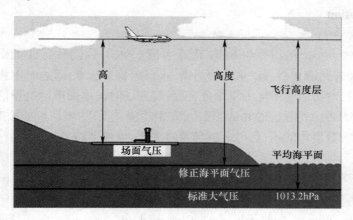

Exercises

Ⅰ. Listen to the following ATIS or VOLMETS, and write down as much information as you can.

1.

Lesson 3-Exercise 1

Unit Two
Preflight to Line-up

2.

3.

4.

5.

Ⅱ. Dictation: Write down what you hear on the recording.
1.
2.
3.
4.
5.
6.
7.
8.
9.
10.
11.
12.

Lesson 3-
Exercise 2

Ⅲ. Translation.

汉译英

1. 山东8592,通知你,滑行道Bravo附近有拖车抛锚,我们马上给你派引导车。

2. 国际1809,跑道幺拐入口被航油污染,入口内移175米,落地时注意。

3. 大韩002,本场通播不能提供服务,准备好离场时向地面申请机场信息。

4. 塔台,邮政9123,滑行道Bravo很滑,左主轮陷在泥里,申请机场援助。

5. 新航109,取消你的飞行计划,机场暂时关闭,风暴即将来临。

英译汉

1. CSN 3518, Harbin Tower, we had an emergency 3 hours ago, the runway had to be foamed, it will be opened in 20 minutes.

2. CPA 576, Wuhan Approach, expect low visibility for your arrival, keep us advised.

3. CSH 2734, runway 04 is partly covered with thin ice, previous EMB145 reported poor braking action on landing 10 minutes ago.

4. It has been snowing for quite a while around here, but there is no report of snow deposit on the runway.

5. Ceiling will rise to 1000 meters soon, airport is expected to be reopened shortly.

Lesson Four
Radio Check and ATC Clearance

Model Exchanges

Radio check

A.

PIL: Hong Kong Ground, CCA 1504, radio check on 121.6, how do you read?

CTL: CCA 1504, Hong Kong Ground, I read you 5.

B.

PIL: Guangzhou Tower, CSN 3309, radio check 121.8.

CTL: CSN 3309, Guangzhou Tower, readability 2, you are broken, adjust your transmitter and give me a short count.

PIL: Roger, CSN 3309.

PIL: Tower, CSN 3309, 1 2 3 4 5, how do you read?

CTL: CSN 3309, I read you 5, loud and clear.

PIL: Thank you, CSN 3309.

ATC Clearance

C.

PIL: Beijing Delivery, CCA 101, request ATC clearance.

CTL: CCA 101, Beijing Delivery, please copy ATC clearance.

D.

PIL: Hong Kong Delivery, CES 5918, 15 minutes before start, destination Hangzhou, information N, request ATC clearance.

CTL: CES 5918, cleared to Hangzhou, D3 departure, flight planned route, initially climb to FL180, request level change en route, squawk 5124, contact Hong Kong Departure on

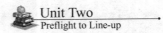
Unit Two
Preflight to Line-up

121.1 when airborne.

PIL: Cleared to Hangzhou, D3 departure, flight planned route, initially climb to FL180, request level change en route, squawk 5124, 121.1, CES 5918.

CTL: CES 5918, read back correct.

E.

PIL: Romeo Tower, AAL 196, request departure information.

CTL: AAL 196, Romeo Tower, surface wind 150 degrees 10 knots, gusting 25, temperature 75 degrees Fahrenheit, dew point 67 degrees Fahrenheit, QNH 1005, departure runway 13.

PIL: Romeo Tower, AAL 196, ready to copy ATC clearance.

CTL: AAL 196, cleared to WBT via flight planned route, N2D departure, turn left after departure, climb and maintain 7000ft, request level change en route, contact 120.26 when airborne, squawk 2514.

PIL: Cleared to WBT via flight planned route, N2D departure, turn left after departure, climb and maintain 7000ft, request level change en route, contact 120.26, squawk 2514, AAL 196.

CTL: AAL 196, correct.

Dialogue

A.

PIL1: Wuhan Ground, CCA 1302, good morning.

CTL: CCA 1302, morning, here is your message, please get ready to copy.

PIL2: Wuhan Ground, … 808.

CTL: Station calling Wuhan Ground, you are blocked, say again your call sign, please speak slower and distinctly.

PIL2: Wuhan Ground, QFA 808, how do you read?

CTL: QFA 808, I read you 5.

PIL1: Wuhan Ground, CCA 1302, we are ready to copy your message.

CTL: CCA 1302, here is your departure information, departure runway 18R, surface wind 330 degrees 8m/s, gusting 15m/s, QNH 1013.

PIL1: Copied, CCA 1302.

PIL3: CSN 3108, request weather information for Chengdu.

CTL: CSN 3108, remain on this frequency, the weather of your destination Chengdu is currently being re-recorded.

PIL1: Wuhan Ground, CCA 1302, 5 minutes before start, gate 2 to Beijing.

CTL: CCA 1302, your ATC clearance.

PIL1: Ready to copy.

Lesson Four
Radio Cheek and ATC Clearance

CTL：CCA 1302, cleared to Beijing via LKO, flight planned route, initial climb to 4,200m, LKO2C Departure, request level change for 10,700m en route, squawk 5310.

PIL1：Cleared to Beijing via LKO, flight planned route, 4,200m initially, LKO2C Departure, request level change for 10,700m en route, squawk 5310, CCA 1302.

CTL：CCA 1302, correct, contact Tower on 121.6, good day.

PIL1：121.6, CCA 1302, good day.

CTL：QFA 808, report number of persons on board.

PIL2：QFA 808, we have 220 persons on board, including six VIPs.

CTL：CSN 3108, here is the latest information for your destination airport.

PIL3：Pass your message, please, CSN 3108.

CTL：There is a 50% chance of thunderstorm and rain at Chengdu Shuangliu Airport.

PIL3：Thanks, CSN 3108.

B.

PIL：Beijing Ground, CSN ···, good morning.

CTL：Station calling Beijing Ground, say again your call sign, your transmission is broken.

PIL：Beijing Ground, CSN 3304, radio check on 121.6, how do you read?

CTL：CSN 3304, Beijing Ground, your signal is unstable, check your transmitter and give me a short count.

PIL：Roger, CSN 3304.

PIL：Ground, CSN 3304, 1, 2, 3, 4, 5, how do you read?

CTL：CSN 3304, read you 3, loud background whistle, check again, and give me a long count.

PIL：Roger, 1, 2, 3, 4, 5; 5, 4, 3, 2, 1, how do you read now?

CTL：CSN 3304, I read you 5, loud and clear.

PIL：Thank you sir, CSN 3304.

New Words and Phrases

adjust *v.* 调整
transmitter *n.* 发射机
surface *n.* 表面,地面,水面
short/long count 短/长数
whistle *n.* (无线电台发出的)尖啸声
block *v.* 阻碍,堵塞,不畅
break *v.* 打断
squawk *n.* , *v.* 应答机编码,把应答机编码设置为……
braking action 刹车效应

Unit Two
Preflight to Line-up

gust *n.* 阵风

gate *n.* 停机门,停机位

start *v.* 开车,开始,是 start-up 的简略讲法

level *n.* 高度(层),此处是 flight level 的简略讲法

request level change en route   航路上申请高度变化

Notes

1. 无线电检查程序。
（1）对方呼号；
（2）本机呼号；
（3）Radio check；
（4）无线电频率；
（5）询问信号质量。

2. 无线电检查回答应按照下列形式。
（1）对方电台呼号；
（2）己方电台呼号；
（3）所发射信号的质量（readability）。

3. 无线电检查对信号质量的描述,见下表。

描述	质　　　量
1	不清楚（Unreadable; I can't read you.）
2	可断续听到（You are cut in and out.）
3	能听清但很困难（Your signal is jammed; your signal is weak.）
4	清楚（Clear）
5	非常清楚（Loud and clear）

4. 放行许可的内容包括目的地、使用跑道、航路飞行规则、航路巡航高度、离场程序、应答机编码,如有必要还应该包括起始高度、离场频率、特殊要求等。驾驶员必须复诵许可内容,在被证实明确无误后,方可申请推出开车。

注：放行许可并非是起飞或进跑道许可。

5. 在没有 ATIS 或者不提供 ATIS 的机场,管制员可按要求向机组提供当时机场的进离场条件（arrival/departure information）。

Exercises

Ⅰ. **Substitution practice.**

A. CCA 1402, cleared to Beijing via Bekol, flight planned route, initial climb to 3,600m, Bekol 2C Departure, request level change for 10,700m en route, squawk 5310.

Lesson Four
Radio Cheek and ATC Clearance

— Tango 1A

—Kilo 7G

— BPK 1G

B. PIL：Xi'an Ground, JAL 076, radio check, 117.3.

CTL：JAL 076, Xi'an Ground, I can't read you, adjust your transmitter.

—you are cut in and out

—your signal is jammed

—your signal is weak

—loud background whistle

Ⅱ. Translation.

汉译英

1. 北京地面,国际1587,无线电检查129.0,你听我几个？

2. 南方3064,听你两个,有很大的背景噪声,调整你的发射机。

3. 上海地面,东方5106,目的地北京,通播 Bravo 收到,准备抄收 ATC 放行许可。

4. 南方3392,可以沿飞行计划航路放行到广州,使用跑道洞拐,沿洞幺标准程序离场,起始高度修正海压九百,修正海压幺洞幺两,巡航高度层九千八,航路上申请高度变化,应答机五幺两四,离地后联系进近幺两幺点幺。

5. 哪个呼叫浦东塔台,重复一遍你的呼号。

英译汉

1. JAL 104, Hong Kong Ground, your signal is unstable, check your transmitter and give me a short count.

2. CSN 3398, you are cleared to Hong Kong via flight planned route, cruising level 10,700m, follow D08 Departure, squawk 3542.

3. KAL 761, Beijing Ground, wind 340 degrees 8m/s, temperature 12 degrees centigrade, dew point 9 degrees centigrade, QNH 1021, departure runway 36L.

4. CES 5128, your read back is incorrect, confirm squawk 6123.

5. I can't read you due receiver failure.

Ⅲ. Fill in the blanks.

1. Guangzhou Tower, CSN 301, _____ Hong Kong, request _____ information.

2. PIL：Mike _____, DYGTV.

CIL：_____(abbreviated call sign), Mike Delivery, go ahead.

3. CCA 1025, _____ to Papa, Tango 2G Departure, filed flight plan route.

4. _____ calling Ground, say again your _____.

5. CSN 3031, I read you 2, _____ your transmitter.

6. Nanjing Tower, CES 5304, _____(无线电检查)121.6, how do you read?

Ⅳ. Listening practice.

1. CSN 3037, Guangzhou Tower, _____ 2, adjust your _____ and give me a short count.

Lesson 4-Exercise 4

Unit Two
Preflight to Line-up

2. Hong Kong Ground, CCA 1604, _____ 121.6, how do you read?

3. Hong Kong Delivery, JAL 739, 15 minutes before start, _____ Beijing with information M, request _____ .

4. JAL 196, Narita Tower, surface wind 120 degrees 8 knots, _____ to 15, temperature 14, _____ 8, QNH 1015, _____ 13.

5. Station calling Ground, _____ your call sign, your transmission is _____ .

Lesson Five
Start-up and Push-back

Lesson 5

Model Exchanges
Push-back
A.
PIL: Apron, DLH 424, stand 5, request push-back.
CTL: DLH 424, push-back approved, facing east.
B.
PIL: Nanjing Tower, JAL 137, Gate 3, request push-back.
CTL: JAL 137, expect 10 minutes delay due to traffic, push back at own discretion.
C.
PIL: Ground crew, ready for push-back.
GND: Confirm brakes released.
PIL: Brakes released.
GND: Commencing push-back.
Start-up
D.
PIL: Beijing Ground, CCA 1358, good morning, ready for start-up.
CTL: CCA 1358, good morning, your slot time is 30, start up 10 minutes before.
PIL: Slot time 30, start up 10 minutes before, CCA 1358.
E.
PIL: Hong Kong Ground, CSN 3306, bay 25, request start-up.
CTL: CSN 3306, start-up approved.
PIL: Start-up approved, CSN 3306.

Unit Two
Preflight to Line-up

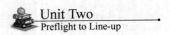

PIL: Ground crew, CSN 3306, ready to start engine number one.

GND: Cleared number one.

PIL: Ready to start number two.

GND: Cleared number two.

PIL: Engine start completed, please disconnect interphone.

GND: Disconnecting.

F.

PIL: Ground, CSN 3392, stand 32, request push-back and start-up.

GND: CSN 3392, push-back approved, stand by for start-up.

Towing procedure

G.

PIL: Beijing Ground, CCA 1878, request tow Air China Boeing 777 from stand A8 to stand 22.

CTL: CCA 1878, Beijing Ground, tow approved from stand A8 to stand 22 via taxiway K and B.

or

CTL: CCA 1878, stand by.

Dialogue

A.

PIL1: Beijing Ground, AFR 725, stand 8, destination Guangzhou, information B, ready for push-back and start-up.

CTL: AFR 725, standby, expect two minutes delay due B737 taxiing behind, call you back.

PIL1: Standing by, AFR 725.

PIL2: Beijing Ground, KAL 410, ready for start-up.

CTL: KAL 410, expect departure 55, start up at your own discretion.

PIL2: Expect departure 55, start up at own discretion, KAL 410.

CTL: AFR 725, push back and start up approved, runway-in-use is 36R.

PIL1: Push-back and star-up approved, runway-in-use 36R, AFR 725.

CTL: Correct, call me back when ready for taxi.

PIL2: (at 45) KAL 410, we wish to delay our start-up, we have to tear down several rows of seats. There is a stretcher case on board. I'll call you back when ready.

CTL: Roger, KAL 410.

PIL3: Beijing Ground, CSH2501, request start-up.

CTL: CSH 2501, standby, I haven't got your flight plan, check with your company, keep us advised.

PIL3: Standing by, CSH 2501.

B.

PIL: Ground crew, cockpit, good evening, ready for push-back.

GND: Good evening, confirm parking brakes released.

PIL: Parking brakes released, please remove the chocks.

GND: Chocks removed, expect push back in 2 minutes.

PIL: Roger, is it possible to start all the engines during push-back?

GND: Affirm.

PIL: Starting sequence 3, 2, 1.

GND: All clear for engine start, starting sequence 3, 2, 1.

PIL: Number 3 rotating.

GND: Number 3 approved.

PIL: Number 2 coming.

GND: Number 2 approved.

GND: Push-back completed, confirm parking brakes set.

PIL: We have set the parking brakes.

GND: Thank you sir.

PIL: Start number 1.

GND: Number 1 approved.

PIL: All three engines have been started, the cockpit instruments show everything normal, please disconnect interphone, thank you and good bye.

GND: Roger, disconnecting, good bye.

New Words and Phrases

stand *n.* 停机位

slot time 预计离场时间

start up 开车

at …own discretion 由……自己掌握

runway-in-use 使用跑道

push-back *n.* 推出(航空器)

stretcher *n.* 担架

brake *n.* 刹车

parking brake 停留刹车

release *n.* 释放,松开,解除

crew *n.* 机组,若干人组成的工作小团体

bay *n.* (停机)泊位

cockpit *n.* 驾驶舱

Unit Two
Preflight to Line-up

commence *v.* 开始

complete *v.* 完成,结束

brake set(停留)刹车设置好

disconnect *v.* 断开,脱离

sequence *n.* 次序,先后

interphone *n.* 内话机

chock *n.* 轮挡

remove/insert *v.* 撤/设置(轮挡)

Notes

1. 开车程序。

(1) 航空器驾驶员申请起动发动机。

航空器位置+申请开车+通播(ATIS 的代码),如:

PIL: Beijing Ground, CSN 3503, stand A5, request start-up, information A.

(2) 空中交通管制员的回答。

A. 同意开车

Start up approved.

B. (预计)开车(时间)分

(expect) start up at (time)

C. 开车时间自己掌握

Start up at own discretion.

D. 预计离场(时间),开车自己掌握

Expect departure (time), start up at own discretion.

E. 你准备好开车了吗?

Are you ready for start-up?

2. 推出程序。

(1) 航空器驾驶员申请推出。

航空器位置+申请推出,如:

PIL: SIA 721, Shanghai Ground, Gate 15, request push-back.

(2) 管制员的回答。

A. 同意推出

Push-back approved.

B. 稍等

Stand by.

C. 推出自己掌握

Push-back at own discretion.

D. 预计(数字)分钟延误(原因)

Expect (number) minutes delay due (reason).

3. 驾驶员推出航空器需征得管制部门的同意,许多机场设有专门的 Apron/Ramp Control(现场指挥)负责此类事务。

4. 用内话机与地面机务联系时可省去航空器呼号。

5. 在繁忙机场,航班延误后,管制单位给出港航空器的预计离场时间称为 slot time。

Exercises

Ⅰ. **Substitution practice.**

A. PIL: Ground, CCA 1581, ready for push-back.

CTL: CCA 1581, expect 10-minute delay due <u>Airbus taxiing behind.</u>

—apron congestion

—vehicle passing by

—sequencing

—ground handling problem

B. PIL: Ground, <u>CSN 3408, bay 25</u>, information D, request start-up.

CTL: <u>CSN 3408, standby for start.</u>

— CCA 1508 gate 5　　　F　　start-up at 35

— CSN 3502 gate 2　　　B　　expect start-up at 35

— JAL 382 bay 20　　　　A　　slot time 40, start up at own discretion

— ACA 761 stand 32　　　C　　expect departure 55, start up at 42

Ⅱ. **Translation.**

汉译英

1. 香港地面,南方 3704,停机位洞拐,通播 Bravo 收到,申请推出开车。

2. 新航 617,预计离场时间四五分,开车时间自己掌握。

3. 地面机务,请证实二发是否转动。

4. 发动机起动结束,请断开内话机。

5. 白鹭 8194,你的预计离场时间由于流量控制推迟 40 分钟,推出开车自己掌握。

英译汉

1. CSN 3503, Beijing Ground, expect 10 minutes delay due traffic.

2. JAL 701, there's a B747 passing behind, I will give you push-back clearance later.

3. UAL 134, stand A9, information C, request start-up for Shenzhen.

4. CSC 8401, expect departure 50, I'll call you back for start.

5. All engines have been started, check everything OK, please remove chocks and other ground equipment.

Ⅲ. **Fill in the blanks.**

1. Ground, CCA 1481, ready for _____ and start-up.

2. CCA 1704, push-back _____, facing west.

Unit Two
Preflight to Line-up

3. DLH 720, confirm brakes _____.

4. CSN 3682, slot time 20, start-up at own _____.

5. CCA 1305, push-back approved, expect runway 36L, D08 _____.

6. Push-back completed, _____ parking brakes set.

7. Ground crew, please confirm number 2 engine is _____.

8. SIA 382, standby, expect 10 minutes _____ due ATC reasons.

9. Engine start completed, please _____ interphone.

10. All _____ for engine start.

IV. Listening practice.

1. Ground, BAW 054, information E, request _____.

2. We intend to _____ our start-up due passenger _____.

3. Beijing Ground, EVA 988, _____ has not _____ yet, we'll be ready in 15 minutes.

4. Tower, we have a _____ delay, could you put me on request for a _____ after 40?

5. Guilin Ground, CCA 1582, the tow bar was broken during _____, we are waiting for a _____.

6. Ground, ready for push-back, _____.

7. All engines have been _____, please remove all _____.

8. CCA 1731, _____ one minute delay due ground _____ problem.

9. CSH 5901, _____, the computer _____ to produce a strip for you. I'll write one out, start-up _____.

10. All clear for _____, starting sequence 4, 3, 2, 1.

11. _____

12. _____

13. _____

14. _____

15. _____

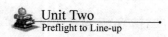
Lesson 5-
Exercise 4

Lesson Six
Taxi and Line-up

Model Exchanges
A.
PIL: Hong Kong Ground, CSN 3056, bay 25, request taxi, information E.

CTL: CSN 3056, taxi via taxiway M, A3 and C to holding point runway 36L, wind 350 degrees 8m/s, QNH 1010.

PIL: Runway 36L, request backtrack from present position and taxiway B4, CSN 3056.

CTL: CSN 3056, backtrack approved and taxi via B4, rest route unchanged.

PIL: Backtrack and use B4, A3 and C to holding point runway 36L, CSN 3056.

CTL: Correct.

B.
PIL: ACA 535, approaching holding point V, request cross runway 19.

CTL: ACA 535, hold short of runway 19.

PIL: Holding short, ACA 535.

CTL: ACA 535, cross runway 19, expedite, report vacated.

PIL: ACA 535, crossing.

PIL: ACA 535, runway vacated.

C.
CTL: BAW 156, hold at holding point A, give way to B747 passing from left to right.

PIL: Holding at holding point A, traffic in sight, BAW 156.

CTL: BAW 156, continue taxi, taxi slower, an aircraft is taxiing in front of you, he is number one departure.

PIL: Roger, taxiing slower, BAW 156.

CTL: BAW 156, caution, A330 overtaking you on your right, he is number 2 departure.
PIL: Roger, BAW 156.
CTL: BAW 156, hold at holding point V, contact Tower 118.7, good day.
PIL: Holding at holding pointV, 118.7, good bye, BAW 156.
D.
PIL: Winton Ground, DLH 169, bay 18, ready for taxi, information G.
CTL: DLH 169, taxi to holding point runway 31 via taxiway E1, N1, N2, and N3.
PIL: Holding point runway 31, request backtrack from present position and taxiway N4.
CTL: DLH 169, negative, traffic taxi in onto taxiway N4.
PIL: Roger, taxiway E1, N1, N2, and N3, DLH 169.
CTL: That's correct.
E.
CTL: CCA 1502, report Airbus on final in sight.
PIL: CCA 1502, traffic in sight.
CTL: CCA 1502, after the landing traffic, line up and wait.
PIL: After the landing traffic, line up and wait, CCA 1502.

Dialogue
A.
PIL1: Wuhan Ground, JAL 101, stand M5, request taxi.
CTL: JAL 101, taxi via taxiway A3 and A.
PIL1: A3 and A, JAL 101.
CTL: JAL 101, taxi with caution, keep close to the centerline when taxiing on taxiway A3.
PIL1: Roger, JAL 101.
PIL2: Wuhan Tower, AFR 188, we'd like to confirm NOTAM bulletin A0013.
CTL: AFR188, NOTAM for your flight is that taxiway A3 is newly repaired on the mid-section of the left side.
PIL2: Roger, thank you.
PIL2: Wuhan Ground, AFR 188, can you change our taxiing route?
CTL: AFR 188, I'd like to confirm your original taxiways and your stand number.
PIL2: We are cleared to taxi via taxiway A3 and A from stand M1, AFR 188.
CTL: AFR 188, backtrack from the present position and use taxiway A4 instead of A3.
PIL2: Backtrack and taxiway A4, AFR 188.
B.
PIL1: Ground, CSH5193, stand B2, request taxi, information L.
CTL: CSH5193, taxi via taxiway E and C to holding point runway 18.

PIL1: Taxiway E and C to holding point runway 18, CSH5193.

CTL: CSH 5193, hold short of the next intersection, give way to the Airbus coming from your left.

PIL1: Hold short of the next intersection, traffic in sight, CSH 5193.

PIL1: CSH 5193, approaching holding point runway 18, request cross runway 18.

CTL: CSH 5193, cross runway 18, report vacated.

PIL1: CSH 5193, crossing runway 18.

PIL1: CSH 5193, runway vacated.

PIL2: Ground, DAL 711, ready for taxi.

CTL: DAL 711, taxi to holding point runway 18 via taxiway B1 and A5.

PIL2: Roger, B1 and A5 to holding point runway 18, DAL 711.

CTL: DAL 711, hold short of A5, give way to B767 taxi in for parking.

PIL2: Holding short, DAL 711.

PIL2: DAL 711 is approaching holding point of runway 18.

CTL: DAL 711, contact Tower 118.1, good day.

PIL2: 118.1, DAL 711, good day.

New Words and Phrases

 taxi *n. v.* 滑行
 taxi via 沿……滑行
 taxiway *n.* 滑行道
 continue *v.* 继续
 hold *v.* 等待
 holding point 等待点
 backtrack *v.* (180°)调头
 cross *v.* 穿越
 hold short of 在……外等待
 vacate *v.* 脱离
 give way to 给……让路
 intersection *n.* 道口,交叉点
 taxi straight ahead 一直往前滑
 follow *v.* 跟在……后面
 traffic *n.* (飞行或航空器)活动,冲突
 in sight 看见,看到
 taxi with caution 滑行时注意
 overtake *v.* 超越
 expedite *v.* 加速,尽快

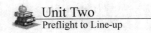

Unit Two
Preflight to Line-up

expedite taxi 加速滑行

taxi slower 减速滑行

take first（second…）left/right 前方第一（第二……）个道口左/右转

take next left/right 下一个道口左/右转

Notes

1. 滑行指令的一项关键内容是许可界限(clearance limit)，许可界限是指航空器按空中交通管制许可之规定可以到达的点。对离场滑行航空器而言，许可界限通常是使用跑道的等待点，但根据当时机场交通状况，它也可以是机场其他任意位置。

2. 航空器欲穿越开放跑道必须获得地面管制许可，并在穿越后报告。

3. 驾驶员在回答 hold 或 hold short 指令时除了用 roger 或 wilco 外，还必须用 holding 或 holding short 作答。

4. 航空器进跑道等待用 line up and wait，而不用 line up and hold。

5. 等待点英文应为 holding point，而不是 holding position。

Exercises

Ⅰ. **Substitution practice.**

A. PIL：Wuhan Ground, DLH 169, bay 18, ready for taxi, information G.

CTL：DLH 169, hold at holding point A, give way to A320 <u>passing left to right</u>.

— coming from your left

— overtaking you on your right

— taxiing in for parking

B. PIL：Beijing Ground, CCA 1503, gate 4, request taxi, information D.

CTL：CCA 1503, taxi to holding point runway 36R, follow <u>DC-10 coming from your left</u>.

— taxi beyond the B737 on your right, then turn left to taxiway C

— taxi behind B757

— taxi slower, caution, taxiways are slippery

Ⅱ. **Translation.**

汉译英

1. 汉莎 076，沿滑行道 Mike 和 Charlie 滑到跑道三六左等待点，滑行时注意。

2. 日航 782，可以穿越跑道两四，脱离报告。

3. 地面管制，我不熟悉本场情况，请派一辆引导车引导我到使用跑道。

4. 达美 655，道口 Victor 外等待，给从右向左穿越的波音 757 让路。

5. 地面，北欧 124，停机位 Delta 两，申请滑到跑道三两的详细滑行指令。

英译汉

1. BAW 024, take next right, follow the aircraft in front of you, I'll keep you advised.

2. JAL 521, stop taxiing immediately, the intersection ahead is blocked by a broken

vehicle.

3. SIA 094, taxi to the north apron via taxiway B3 and A4, stand R3, report marshaller in sight.

4. CCA 1562, caution, B747 overtaking you on your right, he is number one departure.

5. BAW 386, backtrack runway 03, vacate runway via taxiway A1.

III. Fill in the blanks.

1. Hong Kong Ground, GAD, _____ holding point runway 13.

2. JAL 204, _____ at the next intersection, give _____ to DC-10 coming from your left.

3. Ground, BAW 367, _____ from present position.

4. GCD, hold _____ runway 13.

5. Taxi to the holding point A _____ taxiway D2 and B1.

6. GAB, taxi _____ ahead.

7. CDG 7892, after the landing aircraft, _____

8. Continue taxi and I will keep you _____.

9. KAL 075, _____ runway 24, report vacated.

10. JAL 781, _____ crossing runway 14, traffic 4 miles on final.

IV. Listening practice.

1. PIL: Dalian Ground, CSZ 7213, runway _____.
CTL: CSZ 7213, taxi to the _____ via taxiway A and D, the _____ is unserviceable.
PIL: Roger.

2. CCA 1645, after the _____ has passed, line up and _____.

3. CTL: CES 5179, _____ to the left to give way to the A320 taxiing in.
PIL: Pulling in to the left, _____, CES 5179.

4. Hongqiao Ground, DLH 325, _____, stand 12, _____.

5. BFT, taxi _____ until B2, then take the _____.

6. After the B747 passing _____, taxi to the holding point runway 28.

7. Taxi to the _____, cross runway 21 at the _____, report vacated.

8. AFR 105, _____ the _____ line on your left to the _____.

9. KAL 142, make a _____ and _____ to vacate runway via 03 fast turn-off, there's a _____ on the 05 fast turn-off.

10. CHH 7901, taxi with caution, _____ adjacent to stand 27.

Lesson 6-
Exercise 4

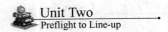

Unit Two
Preflight to Line-up

V. Look at the following pictures and give proper instructions.

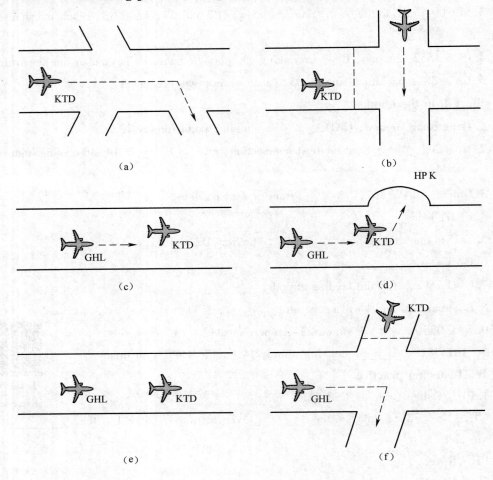

Unit Three

Take-off and Departure

Lesson Seven
Take-off

Lesson 7

Model Exchanges
   A.
   CTL: CSC 7602, report when ready for departure.
   PIL: Wilco, CSC 7602.
   PIL: CSC 7602, ready for departure.
   CTL: CSC 7602, line up runway 18R.
   PIL: Lining up runway 18R, CSC 7602.
   B.
   CTL: CXA 8045, line up and wait, vehicle crossing upwind end of runway.
   PIL: Line up and wait, CXA 8045.
   CTL: CXA 8045, after departure, climb straight ahead until altitude 1200 meters before turning right, cleared for take-off.
   PIL: After departure, climb straight ahead until 1200 meters on QNH before turning right, cleared for take-off, CXA 8045.
   C.
   CTL: JAL 492, runway 22, cleared for take-off, report airborne.
   PIL: Runway 22, cleared for take-off, JAL 492.
   CTL: JAL 492, take off immediately or vacate runway.
   PIL: Taking off, JAL 492.
   PIL: JAL 492, airborne 45.
   CTL: JAL 492, contact Wuhan Approach on 122.75.
   PIL: 122.75, JAL 492.

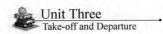
Unit Three
Take-off and Departure

D.

PIL: Urumqi Tower, OKA 7913, ready for departure.

CTL: OKA 7913, Urumqi Tower, wind 190 degrees, 8m/s, cleared for take-off.

PIL: 190 degrees, 8m/s, cleared for take-off, OKA 7913.

CTL: OKA 7913, hold position, cancel take-off, I say again, cancel take-off, unknown vehicle crossing the runway.

PIL: Holding, OKA 7913.

E.

CTL: KAL 481, wind 220 degrees, 10 knots, runway 22, cleared for take-off.

PIL: Taking off, KAL 481.

CTL: KAL 481, stop immediately, KAL 481, stop immediately, fire coming out from wheel well.

PIL: Stopping, KAL 481.

Dialogue

A.

CTL: ACA 096, Beijing Tower, are you ready for departure?

PIL: ACA 096, negative, boarding has not completed yet, we will be ready in 10 minutes, could you put me on request for a slot time after 30?

CTL: ACA 096, approved, call me when ready.

(Later)

PIL: Beijing Tower, ACA 096, ready for departure.

CTL: ACA 096, line up.

PIL: ACA 096, lining up

CTL: ACA 096, wind 340 degrees, 8m/s, runway 36R, cleared for take-off.

PIL: Cleared for take-off, ACA 096.

PIL: Beijing Tower, ACA 096, take-off aborted due to burst tyre.

CTL: ACA 096, vacate the runway via the second fast turn-off on your left.

PIL: ACA 096, negative, we are standing on the stopway, request a tug to tow us back to the apron.

CTL: ACA 096, we are sending you a tug in no time, emergency services will be notified.

B.

PIL1: Beijing Tower, BAW 411, approaching holding point runway 36L.

CTL: BAW 411, hold short of runway 36L, traffic on short final.

PIL1: Holding short, BAW 411.

PIL1: BAW 411, traffic in sight.

CTL: BAW 411, after the landing traffic, line up and wait.

PIL1: Roger, after landing traffic, line up and wait, BAW 411.
(Later)
PIL1: BAW 411, ready for take-off.
CTL: BAW 411, after departure, climb straight ahead until 600m before turning left, surface wind 340 degrees, 8m/s, cleared for take-off.
PIL1: Climb straight ahead until 600m before turning left, cleared for take-off, BAW 411.
CTL: BAW 411, stop immediately, BAW 411, stop immediately, flames coming out from No. 2 engine.
PIL1: Stopping, BAW 411.
PIL2: Beijing Tower, CCA 1308, hold at holding point runway 36L.
CTL: Hold position, emergency traffic on runway.
PIL1: Beijing Tower, BAW 411, we are executing emergency evacuation, request emergency assistance.
CTL: BAW 411, fire truck is on the way, airport ferry and medical service will be available in a few minutes.
…
CTL: CCA 1308, cleared for take-off, report airborne.
PIL2: Taking off, CCA 1308.
PIL2: Airborne 32, CCA 1308.
CTL: CCA 1308, contact Beijing Approach on 129.3, good day.
PIL2: 129.3, good day, CCA 1308.

New Words and Phrases
upwind *n.* 一边,逆风
line up 进跑道
immediate *adj.* 立即的
airborne *adj.* 升空,离地
vacate *v.* 脱离,撤离
burst tyre 爆胎
stopway *n.* 停止道
unknown vehicle 不明车辆
wheel well 轮舱
hold position 原地等待
boarding *n.* 登机
flame *n.* 火焰
execute *v.* 实施,执行

Take-off and Departure

emergency evacuation 紧急撤离
cancel *v.* 取消
climb straight ahead 直线上升

Notes

1. Take off: the action of an aircraft as it becomes airborne. In a more strict sense, the final breaking of contact with the land or water.

Take off 属于指令性管制指挥用语,仅在允许航空器起飞时使用,并要求飞行员复诵。

2. Departure: any aircraft taking off from an airport is referred to as a departure.

其他有关起飞意义的表达用 departure。

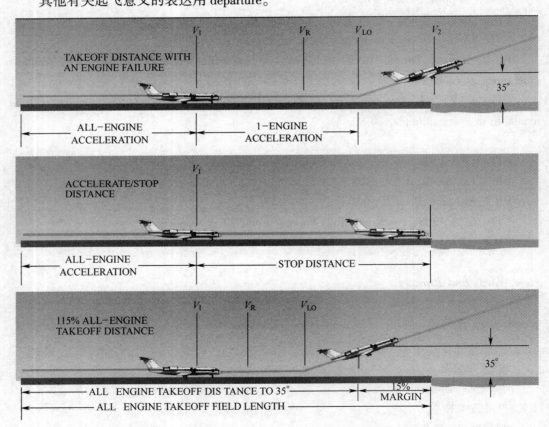

3. 在机场能见度较低的情况下,管制员一般要求机组报告飞机离地的时间。
4. Some of the factors affecting the take-off of aircraft:
（1）Wind
（2）Temperature
（3）Pressure（QNH）
（4）Runway conditions（contamination: rain, snow, ice, low friction…）

(5) Weight of aircraft (fuel, load)
5. Possible reasons for aborting take-off:
(1) Burst tyre
(2) Engine low on power
(3) Undercarriage fire
(4) Nose gear collapsed
(5) Runway incursion
……

Exercises
 I . **Translation.**
英译汉

1. CSZ 8827, climb straight ahead until 900m on QNH before turning right, cleared for take-off.

2. CAL 905, hold short of runway, report MD-11 on final in sight.

3. CDG 9543, cancel, I say again, cancel take-off, workmen on the runway.

4. FDX 528, Guangzhou Tower, take off immediately or hold short of runway, B777 on long final.

5. Nanjing Tower, DLH 198, take-off aborted due to tyre blown out, the right main gear is bogged down, request passenger stairs and airport ferry to take the passengers to the terminal.

汉译英

1. 海南6428,西安塔台,由于军方活动,暂时不能发布起飞指令,稍等进一步指令。

2. 新航512,进跑道,做好立即离场的准备。

3. 荷兰793,跟着短五边的空客380,进跑道等待。

4. 全日空101,原地等待,取消起飞,我重复一遍,取消起飞,有人报告机上有疑似非典病人。

5. 国际1503,起飞后,沿跑道航向上升到修正海压高度九百,然后左转,沿标准程序离场,可以起飞。

Ⅱ. **Fill in the blanks.**

1. MDA 771, are you _____ for _____?

2. After the landing B787 has _____ the runway, line _____ and _____.

3. Cleared for _____, runway 14. _____ immediately or vacate runway.

4. Hold position, _____ take-off.

5. _____ for take-off, wind 230 degrees, 7 knots.

6. ACA 021, report the MD-11 on _____ in _____.

Ⅲ. **Listening practice.**

1. PIL: DAL 711, hold at _____ runway 32.

Lesson 7-
Exercise 3

Unit Three
Take-off and Departure

CTL: DAL 711, line up and _____.

PIL: DAL 711, we have a _____, the _____ seems to be _____.

CTL: Do you require a _____?

PIL: Affirm, request a tug to _____ us back to the apron.

2. PIL: JAL 613, hold at holding point 18L.

CTL: Suggest you hold there _____, the _____ is rapidly approaching the _____ of the runway.

PIL: Wilco, JAL 613.

3. PIL: SIA 509, _____ holding point 29, request return to _____, the _____ are _____.

CTL: Roger, SIA 509, turn into the _____, take the first _____ onto taxiway J.

PIL: Left turn onto taxiway J.

Lesson Eight
Departure

Model Exchanges

A.

CTL: SIA 861, airborne 47, follow G08 departure, continue climb to 600 meters.

PIL: G08, climbing to 600 meters, SIA 861.

CTL: SIA 861, contact Approach 123.6, good night.

PIL: 123.6, SIA 861, good night.

B.

PIL: Hong Kong Departure, DAL 007, passing 3000 feet.

CTL: DAL 007, climb and maintain 6000 feet.

PIL: Climb and maintain 6000 feet, DAL 007.

CTL: DAL 007, continue climb to 7000 feet, report reaching.

PIL: Continue climb to 7000 feet, call you reaching, DAL 007.

PIL: Hong Kong Departure, DAL 007, reaching 7000 feet.

C.

CTL: AFR 309, Hong Kong Departure, turn right heading 130, climb to FL180, expedite until passing FL100.

PIL: Right turn heading 130, climbing to FL180, expediting until passing FL140, AFR 309.

CTL: AFR 309, what is your rate of climb?

PIL: 700 feet per minute, AFR 309.

CTL: Due to traffic, can you adjust your rate of climb to be above FL160 at the FIR boundary?

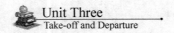

PIL: Above FL160 at the FIR boundary, wilco, AFR 309.

D.

PIL: Departure, AIC 486, passing 900 meters.

CTL: AIC 486, climb to 2100m, expedite until passing 1500 meters.

PIL: 2100m, expediting, call you 1500 meters, AIC 486.

CTL: AIC 486, cancel SID, turn left direct to BTX, climb to 2700 meters.

PIL: Cancel SID, left turn direct to BTX, climbing to 2700 meters, is 3900 meters available, AIC 486?

CTL: AIC 486, negative, can you accept 4500 meters?

PIL: Negative.

CTL: AIC 486, change now to 123.5.

PIL: 123.5, AIC 486, good day.

E.

CTL: SAS 985, Delta Approach, cancel SID, turn left heading 170 to intercept BT 210 radial.

PIL: Cancel SID, left heading 170 to intercept BT 210 radial, SAS 985.

Dialogue

A.

CTL: CCA 1062, report when ready for departure.

PIL: Wilco, CCA 1062.

PIL: CCA 1062, ready for departure.

CTL: CCA 1062, wind 255 degrees, 13 knots, cleared for take-off.

PIL: Cleared for take-off, CCA 1062.

CTL: CCA 1062, airborne 04, climb on present heading to FL110, contact Hong Kong Departure on 128.5.

PIL: Climb to FL110, 128.5, CCA 1062.

PIL: Hong Kong Departure, CCA 1062, good afternoon.

CTL: CCA 1062, Hong Kong Departure, good afternoon, now turn right heading 050 and continue climb to FL170.

PIL: Right turn heading 050 and continue climb to FL170, CCA 1062.

CTL: CCA 1062, now direct to LUF VOR and recleared to FL190.

PIL: Direct to LUF VOR and climb to FL190, CCA 1062.

PIL: Hong Kong Departure, CCA 1062, request FL230.

CTL: CCA 1062, standby, I'll call you back.

CTL: CCA 1062, can you accept FL250?

PIL: Negative.

CTL: CCA 1062, climb to FL230, report reaching.

PIL: Climbing to FL230, CCA 1062.

PIL: Hong Kong Departure, CCA 1062, reaching FL230.

CTL: CCA 1062, roger, change now to Guangzhou Control for higher level, frequency 135.6, good bye.

PIL: 135.6, good bye.

B.

PIL1: Hongqiao Tower, JAL 193, airborne 30.

CTL: JAL 193, climb straight ahead to 1200m, contact Approach on 121.3, good day.

PIL1: Straight ahead to 1200m, 121.3, JAL 193, good day.

PIL2: Hongqiao Tower, CES 5394, ILS approach runway 36R at 1500m.

CTL: CES 5394, Hongqiao Tower, continue approach, report outer marker.

PIL2: Roger, call you outer marker, CES 5394.

PIL3: Hongqiao Tower, KAL 293, hold at holding point runway 36R, request line-up.

CTL: KAL 293, standby, traffic on final.

PIL3: Standing by, KAL 293.

PIL1: PANPAN PANPAN PANPAN, Hongqiao Tower, JAL 193, 22km north at 700m, No. 2 engine loss of power, request return for priority landing.

CTL: JAL 193, Hongqiao Tower, cleared straight-in ILS approach runway 36R, wind 300 degrees, 5m/s, QNH 1001, you are number one.

PIL1: Straight-in ILS approach, runway 36R, JAL 193.

CTL: CES 5394, continue present heading and altitude to overfly the airport, PANPAN in progress.

PIL2: Present heading and altitude to overfly the airport, CES 5394.

PIL1: Hongqiao Tower, JAL 193, we have overrun the runway due weight, we managed to stop on the stopway, request airport assistance.

CTL: JAL 193, roger, we'll send a tug and fire service, any injuries among passengers?

PIL1: Affirm, the flight attendant reported several aged people were injured seriously, JAL 193.

CTL: JAL 193, roger, the ambulance will be with you in a few minutes.

New Words and Phrases

direct *v.* (此处)直飞

climb and maintain 上升到(……高度)保持

rate of climb 上升率,爬升率

intercept *v.* 切,切入

SID: Standard Instrument Departure 标准仪表离场

Unit Three
Take-off and Departure

FIR boundary 飞行情报区边界
available *adj.* 可用的
accept *v.* 接受,飞
outer marker 外指点标,远台
overfly *v.* 飞越
overrun *v.* 超过限度,冲出(跑道)
stopway *n.* 停止道(在可用起飞滑跑距离末端以外地面上一块划定的经过整备的长方形地区)
continue on(此处)保持相同航向
flight attendant 乘务人员

Notes

1. 通常航空器起飞前已经收到 ATC 放行许可,为了保证安全间隔,离场管制员还向 IFR(仪表飞行)航空器发布离场管制指令,指令可以是 SID。

2. 管制员可以根据当时情况取消正在执行的 SID,继而发布新的指令。

3. rate of climb:上升率。如 climb at 10 meters per second,意为上升率 10 米/秒,climb at 1000 feet per minute,意为上升率 1000 英尺/分钟。

4. expedite climb until 1500 meters:尽快上升通过幺五。

5. intercept BT210 radial:切 BT210 度径向线。
intercept 除"切入"意思之外,还有"拦截""截听""截获"等意,如:
（1）We are intercepted by air force…(拦截)
（2）…intercepted urgency call from CAL761…(截听)
（3）glide path intercepted…(截获下滑道信号)

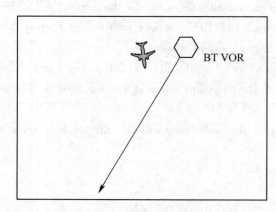

6. continue present heading:保持当前航向。
注:陆空通话中经常使用到"保持",但是需要记住,保持高度和速度一般用 maintain,而保持航向通常用 continue。

Lesson Eight
Departure

Exercises

I. Fill in the blanks.

1. PIL: Baiyun Tower, CSN 3051, ____ Hong Kong, request ____ information.

    CTL: CSN 3051, Baiyun Tower, _____ 18L, wind 160 degrees, 8m/s, ____ 15, temperature 25, dew point 18, ____ 1008.

    PIL: Runway 18L, _____ .

2. PIL: Ground, _____ , _____ .

    CTL: JAL 306, departure runway 16, wind 180 degrees, 08kts, temperature 18, dew point 12, runway is wet, braking action good, QNH 1012.

    PIL: _____ , _____ _____ .

3. PIL: Approach, SIA 404, _____ 4900 feet.

    CTL: SIA 404, _____ climb to FL108, after passing 8900ft, _____ heading to SVA.

4. PIL: Departure, AFR 516, reaching 9800 feet, ____ FL138.

    CTL: AFR 516, ____ to FL138, ____ 131.5 for higher level.

II. Listening practice.

1. CTL: _____ .

    PIL: 126.9, good bye.

    PIL: Foxtrot Control, JAL 905, good morning.

    CTL: JAL905, _____ .

    PIL: JAL 905, 7900ft, heading 090.

    CTL: _____ .

    PIL: Right turn heading 130, climbing to level 217, expediting until passing level 157, JAL 905.

    CTL: _____ .

    PIL: Wilco.

    PIL: JAL 905, reaching FL217.

    CTL: _____ .

    PIL: 128.2, good bye, JAL 905.

2. CTL: _____ .

    PIL: 127.3, good bye.

    PIL: November Control, AAL 108, good evening.

    CTL: _____ .

    PIL: AAL 108, 9800ft, heading 180.

    CTL: AAL 108, _____ .

    PIL: Right turn heading 230, climbing to FL236, AAL108.

    PIL: AAL 108, reaching FL236.

    CTL: _____ .

Lesson 8- Exercise 2

67

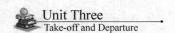
Unit Three
Take-off and Departure

PIL: Climbing to FL276, AAL 108.
CTL: _____.
PIL: 129.5, good bye, AAL 108.

Ⅲ. Substitution practice.

1. PIL: Approach, BAW 711, 900 meters climbing.
CTL: BAW 711, <u>climb straight ahead until 2100 meters before set heading to HLG.</u>
— turn left heading 135 until passing 2100 meters, then direct HLG
— continue climb, direct to HLG
— after reaching 2100m, set heading to HLG
— after passing KR, set heading to HLG

2. PIL: Departure, KAL 306, 2000 feet climbing.
CTL: KAL 306, <u>climb and maintain 8000 feet, report 6000 feet.</u>
— continue climb to 7000 feet, report reaching
— expedite climb until passing 6000 feet
— climb at 1000 feet per minute or less
— cancel SID, continue present heading, climb and maintain 8000 feet
— climb to reach 6000 feet by YW

Ⅳ. Translation.

英译汉

1. CCA 1035, Beijing Approach, climb to reach 5100m by VYK.
2. CES 5287, Hong Kong Approach, climb at 15 meters per second or greater.
3. Xi'an Approach, CSN 3579, we have justencountered severe turbulence, what kind of traffic is there ahead of us?
4. DAL 057, due to traffic, can you adjust your rate of climb to be above FL190 at the FIR boundary?
5. Chengdu Approach, JAL 762, we are returning, we seem to have wheel well fire, the warning light has just flashed on, request priority landing and emergency services.

汉译英

1. 长荣813,左转航向245切入HL VOR 270度径向线。
2. 立荣639,在LIG之前上升到八千四,尽快上升通过拐两。
3. 国泰799,取消SID,右转直飞KDL,上升到六千三保持。
4. 英航134,加入Bravo 09标准离场程序,上升到拐两,通过六千三报告。
5. 白鹭9816,立即左转避让不明冲突航空器。

Unit Four

En-route

Lesson Nine
Level Information and Position Report

Model Exchanges

A.

PIL: Tango Control, CCA 1052, good morning.

CTL: CCA 1052, Tango Control, go ahead.

PIL: CCA 1052, YVR at 35 at 10,100 meters, estimating KOM at 50, next UHW.

CTL: CCA 1052, roger, next report at KOM.

PIL: Wilco, CCA 1052.

B.

PIL: CSZ 7614, 10,400 meters, request descent.

CTL: CSZ 7614, maintain 10,400 meters until further advised.

PIL: Maintaining, CSZ 7614.

CTL: CSZ 7614, descend to 8400 meters, report passing even levels.

PIL: Descending to 8400 meters, call you passing even levels, CSZ 7614.

PIL: CSZ 7614, passing 9800 meters.

C.

CTL: QFA 235, climb when ready to FL190.

PIL: Roger, climb to FL190, QFA 235.

PIL: QFA 235, leaving FL150 for FL190.

(Later)

PIL: QFA 235, approaching FL190.

CTL: QFA 235, continue climb to FL230, expedite until passing FL210.

PIL: QFA 235, unable expedite climb due weight.

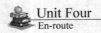
Unit Four
En-route

D.

CTL: JAL 842, omit position report on this frequency.
PIL: Wilco, JAL 842.
CTL: JAL 842, resume position report.
PIL: Wilco, JAL 842.
CTL: JAL 842, next report at Colinton.
PIL: Wilco, JAL 842.
PIL: JAL 842, Colinton 47, FL180, descending to FL140, abeam KTN at 55.
CTL: JAL 842, report 25 miles DME from Kennington.
PIL: Wilco, JAL 842.
CTL: JAL 842, report your DME distance from Kennington.
PIL: 26 miles, JAL 842.
CTL: JAL 842, report passing 270 radial Kennington VOR.
PIL: Wilco, JAL 842.

Dialogue

A.

PIL: Beijing Control, BAW 375, good afternoon.
CTL: BAW 375, go ahead.
PIL: BAW 375, 8400 meters, heading 130.
CTL: BAW 375, continue present heading, climb to 9200 meters on standard, report reaching.
PIL: Climbing to 9200 meters, call you reaching, BAW 375.
PIL: BAW 375, level at 9200 meters.
CTL: BAW 375, continue climb to 9800 meters.
PIL: Climbing to 9800 meters, is 10400 meters available?
CTL: Negative, can you accept 11000 meters?
PIL: Negative, we can't manage due to weight.
CTL: Change now to 129.2.
PIL: 129.0, BAW 375.
CTL: Negative, change to 129.2, I say again, 129.2.
PIL: 129.2, BAW 375.

B.

PIL: Alpha Control, ACA 535, estimating SUR 25, 9800m maintaining.
CTL: ACA 535, descend and maintain 6600m, report passing 8400m.
PIL: Descending to 6600m, report passing 8400m, ACA 535.
PIL: ACA 535, passing 8400m.

72

Lesson Nine
Level Information and Position Report

CTL: ACA 535, roger.

CTL: ACA 535, advise if able to cross SUR at 15.

PIL: ACA 535, affirm.

CTL: ACA 535, cross SUR at 15 or before.

PIL: Cross SUR at 15 or before, ACA 535.

PIL: ACA 535, approaching 6600m, request 6000m.

CTL: ACA 535, negative for the moment due traffic, expect lower level after SUR.

PIL: ACA 535.

CTL: ACA 535, report position, you didn't report over SUR which is a compulsory reporting point.

PIL: ACA 535, we didn't realize we'd passed SUR, there must be something wrong with our FMS, request further clearance.

CTL: ACA 535, continue descend to 5400 meters, VFR, contact Alpha Approach 118.5.

PIL: Down to 5400 meters, 118.5 for Alpha Approach, ACA 535.

PIL: Alpha Approach, ACA 535, request visual approach due to compass failure.

CTL: ACA 535, cleared for visual approach, report runway 36L in sight.

PIL: Wilco, ACA 535.

New Words and Phrases

abeam *adv.* 正切

DME 测距仪

VOR 全向信标

distance *n.* 距离

maintain *v.* 保持(高度或速度)

continue *v.* 保持(航向)

leave *v.* 离开

level *n.* 高度(层); *v.* 在(……高度)改平

omit *v.* 省略

compass *n.* 罗盘, 罗盘仪

radial *n.* (VOR)径向线

resume *v.* 恢复

position report 位置报告

heading *n.* 航向

flight level 飞行高度(层)

CRP(Compulsory Reporting Point) 强制报告点

FMS(Flight Management System) 飞行管理系统

Unit Four
En-route

Notes

1. 国际民航组织（ICAO）推荐下列格式为航路上航空器位置报告程序：

（1）本机呼号

（2）本机当前位置

（3）当前时间

（4）当前高度

（5）下一位置以及预计到达时间

（6）再往后的重要点

2. 在区域管制中除了驾驶员按照规定报告位置外，管制员还根据当时条件要求驾驶员进行附加位置报告，包含：

（1）通过某一航路点报告，如：

CDG 8815, report VYK.

（2）通过 DME 的某一距离报告，如：

CDG 8815, report 30 VYK DME.

（3）报告距 DME 台的距离，如：

CDG 8815, report distance from VYK.

（4）通过某一 VOR 径向线报告，如：

CDG 8815, report passing 165 radial VYK VOR.

（5）在某一 VOR 径向线上以及距离报告，如：

CDG 8815, report 30 DME 165 radial VYK VOR.

3. 高度信息。

（1）Maintain (level) to… 保持（高度）到……

（2）Maintain (level) until passing… 保持高度过……

（3）Maintain (level) until (minutes) after passing…过……之后保持（高度）直到……分

（4）Maintain (level) until advised by (name of unit)…保持（高度），等待……进一步通知

（5）Maintain (level) until further advised 保持（高度），等待进一步通知

（6）Report passing odd/even levels 过奇/偶数高度层报告

4. 高度信息主要涉及上升及下降术语，下面几个词是常用的高度报告术语，示例见下图。

（1）Leaving　　现在离开

（2）Reaching　　现在到达

（3）Passing　　现在通过

（4）Maintaining　　正在保持

（5）Approaching　　正在接近

5. Factors affecting the aircraft en route:

Lesson Nine
Level Information and Position Report

- Passing 8400m
- Leaving 8400m
- Maintaining 8400m
- Reaching 8400m
- Approaching 8400m

(1) Temperature
(2) Pressure
(3) Weight
...

Exercises

I. Substitution practice.

A. CTL: ACA 156, <u>report passing 6300 meters</u>.
PIL: Wilco, ACA 156.
— next report at VYK
— omit position report until VYK
— report intercepting the 075 radial of VYK VOR
— report 25 miles from VYK DME
— resume position reporting after VYK
— report passing the 145 radial of VYK VOR

B. PIL: Shanghai Control, CDG 7514, <u>level at 8400 meters</u>.
   CTL: CDG 7514, roger.
— leaving 8400 meters for 10400 meters
— reaching 8400 meters, request higher level.
— passing 8400 meters
— maintaining 8400 meters
— approaching 8400 meters, request descend to 7200 meters

Lesson 9-Exercise 2

II. Listening practice.

1. PIL: MAYDAY MAYDAY MAYDAY, JAL 711, there is _____, we are making an _____ to 3000 meters, heading to Hongqiao Airport for _____.

Unit Four
En-route
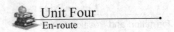

2. PIL: Alpha Control, KLM 902, LGH 13, 10,100 meters, _____.

CTL: KLM 902, roger, report _____.

CTL: KLM 902, maintain present level, be aware of _____, keep me advised of your intention.

…

PIL: MAYDAY MAYDAY MAYDAY, KLM 902, _____, we heard a loud explosion, seems we are losing control.

3. PIL: Delta Control, OKA 8619, we just got _____, 5 passengers have been injured, we will have to divert to YM, request _____.

CTL: OKA 8619, Delta Control, report your position.

PIL: We are about _____ at 9200 meters, OKA 8619.

CTL: OKA 8619, roger, you are cleared to YM, how many _____ will be needed on landing?

PIL: The cabin crew says 5 passengers will have to be taken to the hospital and there are 10 others _____, OKA 8619.

Ⅲ. Translation.
英译汉

1. Wuhan Control, CES 5183, KVN 45, 10,400 meters, estimating HID 1502, next HDE.

2. ANA 013, direct to OLR VOR, maintain present flight level until OLR VOR, report passing HIG 275 radial.

3. ACA 913, maintain 8400 meters, expect higher in 20 kilometers.

4. Shenyang Control, CSC 7182, 7200 meters maintaining, our fuel endurance is only 45 minutes, we'd like to refuel at Shenyang.

5. Urumqi Control, CSN 3375, our leading edge is iced over, we have trouble maintaining level, request immediate descent.

汉译英

1. 法航 317, 停止上升, 在 DNM 上空有会聚飞行, 保持九千二, 过 DNM 报告。

2. Bravo 区域, 美联航 191, 过 SMK, 预计飞越 RK 洞拐分, 高度幺洞四保持, 申请下降。

3. 白鹭 6198, 武汉区域, 保持速度不大于五百五十公里小时直到通过 WG。

4. PANPAN PANPAN PANPAN, Sierra 区域, 英航 612, 刚刚截听到美利坚 096 的紧急讯号, 一位女乘客即将分娩, 申请备降 Sierra 机场, 该航班的位置在 HGR 以南 45 海里, 高度层两九洞。

5. Foxtrot 区域, 北欧 814, BCP 以西 40 海里, 高度层三幺洞, 右发熄火, 正在排除故障。

Lesson Ten
Join, Cross and Leave Airways

Model Exchanges

A.

PIL: KAL 982, request clearance to join B473 at 11600m.

CTL: KAL 982, 11600m not available due to traffic congestion, alternative is 12200m, advise.

PIL: KAL 982, accept 12200m.

B.

PIL: Foxtrot Control, JAL 018, request clearance to join V21 at LPY.

CTL: JAL 018, Foxtrot Control, report position.

PIL: JAL 018, passing BTU heading 150, FL160, request join airway V21 at LPY.

CTL: JAL 018, Foxtrot Control, cleared to join V21 at LPY before 1538, FL160, maintain own separation, report entering the airway.

C.

PIL: BAW 018, request crossing V12 at NEF.

CTL: BAW 018, pass your message.

PIL: BAW 018, Cessna 172, 35km east of NEF at 9800 meters, estimating NEF 30, request clearance to cross V12 at NEF, 9800 meters.

CTL: BAW 018, cleared to cross V12 at NEF before 32, maintain 9800 meters.

D.

PIL: AFR 511, request clearance to leave controlled airspace, northeast of PWR at 8100m.

CTL: AFR 511, cleared to leave controlled airspace via PAG, maintain 8100m while in

77

Unit Four
En-route

controlled area.

PIL: Tango Control, AFR 511, unable to leave controlled area at 8100m due to equipment.

CTL: AFR 511, roger, can you accept 7800m?

PIL: AFR 511, affirm, leaving controlled area via PAG at 7800m.

Dialogue

A.

PIL1: CSN 3342, WG, estimating SH 41, request clearance to join A461 at SH.

CTL: CSN 3342, cleared to Guangzhou, flight planned route, 9200m, join A461 at SH at 9200m.

PIL1: Cleared to Guangzhou, flight planned route, 9200m, join A461 at SH at 9200m to enter controlled airspace, CSN 3342.

CTL: CCA 1308, maintain present altitude, be aware of the weather ahead of you, keep me advised of your intention.

PIL2: Roger, maintaining present level, call you back, CCA 1308.

PIL1: MAYDAY MAYDAY MAYDAY, CSN 3342, severe vibration on the left side, we heard loud explosion, seems we are losing control.

CTL: CSN 3342, roger MAYDAY, what's your intention?

PIL1: CSN 3342, we're sorting out what is going on. I will call you back.

CTL: CSN 3342, roger.

CTL: CCA 1308, what is your flight condition?

PIL2: Wuhan Control, CCA 1308, we've just run into severe turbulence and we would like to turn right to detour the weather.

CTL: CCA 1308, negative to turn right due to emergency traffic, can you accept 8400m?

PIL2: CCA 1308, affirm for 8400m.

CTL: CCA 1308, call me back when clear of the weather.

PIL2: CCA 1308, wilco.

PIL1: Wuhan Control, CSN 3342, we have lost No. 2 engine, we are trying to continue the flight by the remaining engine, request priority landing at Wuhan.

CTL: CSN 3342, roger, continue present heading and down to 7800m, contact Wuhan Approach on 118.1.

PIL1: 7800m and contact Approach 118.1, CSN 3342.

PIL2: Wuhan Control, CCA 1308, clear of weather.

CTL: CCA 1308, maintain present level.

Lesson Ten
Join, Cross and Leave Airways

PIL2: Maintaining, CCA 1308.

B.

PIL: Guilin Control, CES 5402, over LLC at 10100m to Guangzhou, estimating UIT at 51, next KKD, request clearance to cross airway R434 at YW.

CTL: CES 5402, cleared to cross airway R434 at YW at 10100m.

PIL: Cleared to cross airway R434 at YW, 10100m, CES 5402.

CTL: CES 5402, report YW.

PIL: Wilco, CES 5402.

CTL: CES 5402, Guilin Control, it looks like the flight conditions are getting pretty rough in Guilin area at the moment. Over Guilin now the weather is really harsh. I have got some reports coming on the way, standby.

PIL: CES 5402.

(Later)

CTL: CES 5402, there is a chance of CBs beyond TU north of your course.

PIL: CES 5402, roger, we are on top of a solid overcast; the tops are about 10200 meters increasing towards the north and the west.

CTL: CES 5402, what is the flight condition? Do the tops look smooth?

PIL: Air is very smooth and the tops are extremely stratified, CES 5402.

CTL: CES 5402, thank you.

New Words and Phrases

 congestion *n.* 拥挤,拥塞
 alternative *n.* 备份,替换物
 separation *n.* 间隔,分隔,间距
 leading edge 机翼或螺旋桨的前缘
 join *v.* 加入
 cross *v.* 穿越
 enter *v.* 进入
 airway *n.* 空中航线,航路
 airspace *n.* 空域
 vibration *n.* 震动,颤动
 explosion *n.* 爆炸,爆发
 sort out 清理,解决
 weather *n.* 天气(尤指不利天气)
 priority landing 优先着陆
 rough *adj.* 粗暴的,严峻的,剧烈的
 harsh *adj.* 严酷的,严峻的

Unit Four
En-route

on top of 在……之上
smooth *adj.* 平稳的,流畅的
stratify *v.* (使)分层,成层

Notes

1. 若航空器驾驶员要求的飞行高度层已被占用(occupied),管制员会提供可用高度层。

2. 穿越航路的航空器应在穿越之前向相关管制部门提出申请,并同时报告自己的位置及穿越的航路代号,管制员应同时发布其他相关指令及保持必要的飞行间隔。

3. 通常管制员会提供一指定点、指定高度层用以使相关航空器飞离管制区域。

4. KAL 982, request clearance to join B473 at 11600m. B473 是航路代号,类似的航路代号还有:

　　A——Amber,南北向主航路
　　B——Blue,南北向辅航路
　　G——Green,东西向主航路
　　R——Red,东西向辅航路
　　J——JET,高空航路(美国)
　　U——UPPER,高空航空(欧洲)
　　K——Kopter,直升机航路
　　V——Victor,(美)中低空航路
　　HL——High Level,高高度航路
　　GR——Gulf Route,海湾航路
　　ADR——Advisory Rout,咨询航路
　　Dom——Domestic,国内航路
　　D——Direct,直飞航路
　　NAT——北大西洋航路
　　OTR——海洋过渡线
　　POR——区域导航航线
　　SD——SuperSonic,超声速航线

Exercises

Ⅰ. **Fill in the blanks.**

1. Wuhan Control, SAS 108, _____ MDK 34, 7800m descending.

2. GBT, cleared to leave _____ area via HOG, _____ FL290 while in controlled area.

3. Kilo Control, AFR 718, _____ JUK 4200m at 09, _____ CBM at 55.

4. CSN 3310, _____ for level change at the moment, _____ higher after LM.

Lesson Ten
Join, Cross and Leave Airways

5. CCA 1582, landing delays at Beijing, can you _____ time en route?

6. ACA 324, descend immediately to 6000m _____ traffic.

7. DAL 303, _____ to MK and hold at 9200m due to delaying action, standard _____.

8. SAS 996, flight planned route, 8400m, _____ R330 at KFI at 8400m.

9. JAL 305, 8900m is not _____, can you _____ 9500m?

10. Wuhan Control, CES 5321, HIV 45, 12500m, ETO EID 58, _____ IDO.

II. Listening practice.

1. PIL: ANA 342, request clearance to _____ controlled airspace _____ of KDG at FL _____.

CTL: ANA 342 cleared from _____ miles northwest of KDG to PRL via B12, _____ FL _____.

2. PIL: DSE, request _____ to leave controlled airspace by descent.

CTL: DSE, cleared to _____ controlled airspace by descent, report _____ 5900 feet on QNH1010.

3. PIL: MAYDAY MAYDAY MAYDAY, AFR 188, we have lost number one _____ and the other is also losing _____.

4. CTL: Descend to any _____ you like and keep me _____.

5. PIL: Wuhan Control, CES 5321, _____ at GOSMA, ETO LKO 2103, 8900m _____, request descent.

6. PIL: Tango Control, CCA 1879, _____ JFM, heading 290 at FL118, VMC to cross _____ at _____.

7. PIL: Alpha Control, UAL 550, request _____ clearance at KG.

8. PIL: Guangzhou Control, CSC 8134, we have just been caught in _____, about 15 passengers have been hurt, request medical _____ on landing.

9. PIL: Alpha Control, ACA 292.

CTL: _____.

PIL: ACA 292, request clearance to join G12 at ROD.

CTL: _____.

PIL: Cleared to B via ROD, flight planned route, FL276, to enter controlled airspace FL276, ACA 292.

CTL: _____.

10. PIL: Charilie Control, DLH 721.

CTL: _____.

PIL: DLH 721, B737, 30 miles northeast of HGT, FL291, HGT 42, request clearance to cross Upper Red 207 at HGT.

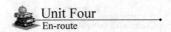
Unit Four
En-route

CTL: _____ .

PIL: Cleared to cross Upper Red 207 at HGT, FL291, DLH 721.

CTL: _____ .

PIL: DLH 721.

III. Translation.

英译汉

1. CSN 3392, join route R200 at BEBEM at 7800 meters.

2. CSN 3392, cross route H25 at WYN at 9800 meters at 55 due traffic.

3. CES2180, cleared to leave controlled airspace via KTD, maintain 8400m while in controlled area.

4. N231, 10 miles south of MK, heading 210, passing9000ft, estimating PS at 30, request cross V538 at PS.

5. CCA1518, hold at VW at 10100m, expect cross clearance at 05.

汉译英

1. 武汉区域,东方5321,管制边界GOSMA,预计飞越LKO洞两分,高度幺洞四保持,申请穿越边界。

2. 南方3307,由于高度九千五有相对活动,不能同意高度幺洞幺,预计相遇时间幺三分。

3. 新西兰538,波音767,高度幺洞四,距ZHO八十公里,申请在ZHO穿越航路V21。

4. 高度幺洞幺不可用,由于冲突,能否接受高度幺洞拐?

5. 日航781,在XS上空等待,保持高度五幺,白云机场交通拥挤,预计进一步许可两洞分。

Lesson Eleven
ADS, RVSM, En-route Holding and Descent

Model Exchanges
A.
CTL: CSC 9173, advise ADS-B capabilities.
PIL: ADS-B transmitter 1090 data link, CSC 9173.
B.
CTL: CES 5931, transmit ADS-B ident.
PIL: Roger, ident.
CTL: CES 5931, identified.
(Later)
CTL: CES 5931, re-enter ADS-B aircraft identification.
PIL: Wilco/Unable, CES 5931.
C.
CTL: CES 5427, Shanghai Control, ADS-B out of service, descend to 10100 meters, direct to Guangzhou.
PIL: Descending to 10100 meters, direct to Guangzhou, CES 5427.
D.
CTL: CSN 3246, confirm RVSM approved.
PIL: Affirm RVSM, CSN 3246.
E.
CTL: SIA 786, unable issue clearance into RVSM airspace, maintain 4800m.
PIL: Wilco, 4800m maintaining, SIA 786.

Unit Four
En-route

F.

PIL: JAL 384, unable RVSM due turbulence.

CTL: Roger, report able to resume RVSM.

G.

PIL: Wuhan Approach, CSN 3507, request descent.

CTL: CSN 3507, Wuhan Approach, expedite descent to 1800 meters on QNH 1018, maintain own separation and VMC from 4200 meters to 2400 meters.

PIL: Leaving 5400 meters for 1800 meters on QNH 1018, maintain own separation from 4200 meters to 2400 meters, CSN 3507.

H.

CTL: B6814, Nanjing Approach, hold visual between the Nanjing Yangtze River Bridge and the Purple Mountain.

I.

CTL: CES 5310, hold on the 210 radial of the WB VOR between 25 miles and 30 miles DME at 2100m, inbound track 030 degrees, left hand pattern, expect approach time 32.

PIL: Hold on the 210 radial of the WB VOR 25 miles and 30 miles DME, 2100m, inbound track 030 degrees, left turns, CES 5310.

J.

CTL: CCA 1982, Juliett Approach, maintain 5100m, hold over GKL as published, traffic congestion at Juliett airport, expect further clearance at 1915.

PIL: Roger, hold over GKL, 5100m, CCA 1982.

PIL: Guangzhou Approach, CCA 1982, request extended holding.

CTL: CCA 1982, hold between CK and PU at 4800m, right hand pattern, expect further clearance at 25.

PIL: Hold between CK and PU at 4800m, right hand pattern, CCA 1982.

K.

CTL: SAS 996, hold at PM at 8900m.

PIL: Hold at PM at 8900m, what is the delay? SAS 996.

CTL: SAS 996, delay not determined, expect further clearance at 1805.

PIL: Roger, SAS 996.

PIL: Beijing Control, SAS 996, our leading edges are iced over, we can not hold at present level, request immediate descent.

CTL: SAS 996, descend and maintain 6900m.

Dialogue

A.

PIL1: Wuhan Control, CES 5402, maintaining 4500m on standard, passing XSH 1412,

Lesson Eleven
ADS, RVSM, En-route Holding and Descent

request climb.

CTL: CES 5402, Wuhan Control, climb to 10100m on standard.

PIL1: Climb to 10100m on standard, CES 5402.

PIL2: Wuhan Control, CCA 1307, passing DM, ETO LKO 1421, 8900m maintaining, request descent.

CTL: CCA 1307, descend to 6900m, report your ETA of Wuhan.

PIL2: Descending to 6900m, ETA Wuhan 29, CCA 1307.

CTL: CCA 1307, confirm you can lose 6 minutes en-route. There will be a delay due to ATC reasons.

PIL2: Negative, CCA 1307.

CTL: CCA 1307, maintain 6900m, expect hold over LKO.

PIL2: Roger, maintaining 6900m and request holding instructions, CCA1307.

CTL: CCA 1307, hold on the 205 radial of the LKO VOR between 60 and 70km DME, 6900m, inbound track 025, left hand pattern, expect further clearance at 30.

PIL2: Hold on the 205 radial of the LKO VOR between 60 and 70km DME, 6900m, inbound track 025, left hand pattern, CCA 1307.

PIL3: Wuhan Control, BAW 141, 8100m maintaining, we have encountered some turbulence, request 10100m.

CTL: BAW 141, Wuhan Control, unable to approve 10100m, 9500m is available.

PIL3: Accept 9500m, BAW 141.

CTL: BAW 141, climb to 9500m.

PIL3: Climbing to 9500m, BAW 141.

PIL4: Wuhan Control, JAL 785, just now, a male passenger attempted to enter the cockpit, it seems that he has taken some drugs, while being escorted back to his seat, the passenger attacked other passengers.

CTL: JAL 785, Wuhan Control, copied. What is the situation now? Are there any passenger injures or damage to the aircraft?

PIL4: There is no damage to the aircraft and there are no injured passengers, the attacker has now become unconscious, request emergency landing at Wuhan airport, JAL 785.

CTL: JAL 785, expedite descent to 4800m, do you need any medical assistance?

PIL4: Descending to 4800m, we need an ambulance and a physician upon arrival, JAL 785.

CTL: JAL 785, an ambulance and a physician will be ready.

B.

PIL1: Wuhan Control, CES 5402, P51 1210 at 8400m, estimating XSH 18, request descent.

CTL: CES 5402, Wuhan Control, descend to 6600m on standard, report reaching.

Unit Four
En-route

PIL1: Roger, descend to 6600m, CES 5402.

PIL2: Wuhan Control, CSC 8892, passing ZF 1212, 9200m, request cruising level 10400m.

CTL: CSC 8892, maintain 9200m, opposite traffic, B737 at 9500m, estimated passing time 15, report traffic in sight and clear of traffic.

PIL2: Roger, maintaining 9200m, CSC 8892.

PIL2: Wuhan control, CSC 8892, the Boeing 737 is in sight and clear of traffic, request climb.

CTL: CSC 8892, maintain 9200m, I will call you back.

PIL2: Roger, 9200m, CSC 8892.

CTL: CSC 8892, climb to 10400m, report reaching.

PIL2: Climbing to 10400m, call you reaching, CSC 8892.

PIL1: Wuhan Control, CES 5402, reaching 6600m.

CTL: CES 5402, Wuhan Control, I have just got the current weather information of Wuhan airport: the CBs are building up at the airport, the reported ceiling is 300m, visibility is 500m, and it is deteriorating, meanwhile, there is a windshear warning at 400m on final of runway 04.

PIL1: CES 5402, roger, we decided to proceed to Yichang airport which is our alternate, what is the weather of Yichang airport?

CTL: Standby, I'll check for you.

Words and Phrases

published *adj.* 公布的
extend *v.* 延伸,扩展,扩大
inbound/outbound *n.* 入航/出航,入航边/出航边
track *n.* 航迹
standard *adj.* 标准的
ETO(Estimated Time Over) 预计飞越时间
ETA(Estimated Time of Arrival) 预计到达时间
cockpit *n.* 驾驶舱
escort *v.* 护送,护卫
unconscious *adj.* 无意识,失去知觉的
assistance *n.* 帮助,援助
physician *n.* 内科医生
CB *n.* 雷暴
ceiling *n.* 云幕,云底高
deteriorate *v.* 恶化

alternate *n.* 备降机场

Notes

1. ADS（Automatic Dependent Surveillance）自动相关监视：飞机通过数据链自动发送和接收机载设备所提取的监控信息，如识别位置（高度、经度、纬度）、速度及意向信息等。

2. RVSM（Reduced Vertical Separation Minimum）缩小垂直间隔：即将现代喷气式民航客机巡航阶段所在用的飞行高度层 8900 米至 12500 米（含）之间的垂直间隔标准由 600 米缩小到 300 米，从而增加空域容量，提高航空公司的运行效益，减轻空中交通管制指挥的工作负荷。

巡航高度层（东单西双）：

（1）航空器进行航路和航线飞行时，应当按照所配备的巡航飞行高度层飞行。

（2）真航线角在 0°～179°范围内的，巡航高度层按照下列方法划分，示例见下图：

高度由 900 米至 8100 米，每隔 600 米为一个高度层；

高度在 8900 米至 12500 米，每隔 600 米为一个高度层；

高度在 12500 米以上，每隔 1200 米为一个高度层。

（3）真航线角在 180°～359°范围内的，巡航高度层按照下列方法划分：

高度由 600 米至 8400 米，每隔 600 米为一个高度层；

高度在 9200 米至 12200 米，每隔 600 米为一个高度层；

高度在 13100 米以上，每隔 1200 米为一个高度层。

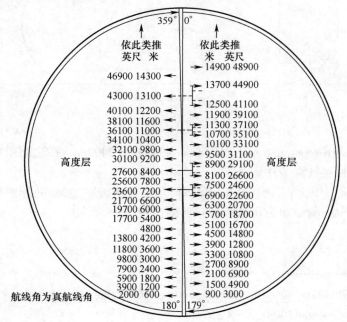

飞行高度层配备标准示意图

Unit Four
En-route

3. 下降指令应复诵,复诵时还应讲明本机当前高度状况,比如当前本机高度或正离开某一高度向指定高度下降等。

4. 下降指令常常有其他条件。

(1) 地点限制:after passing VKY descend to 4200m

(2) 高度限制:stop descent at 6000m

(3) 时间限制:expect further descent in 10 minutes

(4) 下降时调整下降率:descend to 1200m, expedite until passing 2400m

5. 通常情况下等待程序都有明确规定,但驾驶员在准备不充分或欲进一步证实时可要求管制员详细说明等待程序。

6. 按国际民航组织规定,管制员发布的等待指令应包括以下内容并按以下顺序发布,标准等待模式示意见下图:

(1) fix(等待点)

(2) level(高度)

(3) inbound track(入航航迹)

(4) right or left turn(右或左转弯)

(5) time of leg(出航时间)

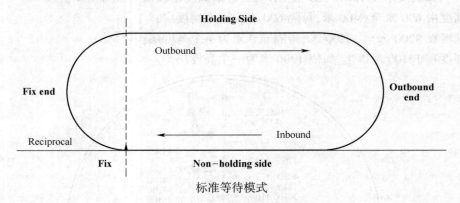

标准等待模式

7. maintain own separation and VMC from 4200m to 2400m. 在高度四两和两千四之间自行保持间隔,目视气象条件。

8. 在程序管制中,遇有两机相对活动时,若用目视解除冲突,必须两机都报"飞机过去了/冲突解除"后,管制员方可继续发布后续指令。

Exercises

I. Substitution practice.

A. CTL:CSN 3576, <u>descend immediately to 4200m due traffic.</u>

PIL:<u>Descending to 4200m</u>, CSN 3576.

— descend to 2100m at CH due delay

— descend to 6000 feet before LK for spacing

Lesson Eleven
ADS, RVSM, En-route Holding and Descent

— descend to 3600m on QNH 1008 for spacing

— descend and maintain 4800m until passing LM due delay

B. PIL: Delta Approach, ACA 321, 6900m, CM 15.

CTL: ACA 321, descend to 5700m, report CM.

PIL: Leaving 6800m for 5700m, call you CM, ACA321

呼号	现在高度	预计下降点及时间	管制员下降指令
BAW 018	7200m	DM 50	4800m
JAL 434	6600m	SU 05	4200m
AAL 626	4800m	NG 15	2400m
SIA 717	6300m	PG 46	2700m

II. Fill in the blanks.

1. CTL: CCA 1582, descend when_____ to 2400m, QNH1012.

PTL: Descending to 2400m on QNH 1012, CCA 1582.

2. PIL: JAL 741,_____ ZFK at 05,_____ S.

CTL: JAL 741, descend to 2700m on QNH 1011,_____ passing odd_____.

3. CTL: CES 5102, descend to 1200m,_____ until passing 2400m.

PTL:_____ 6000m, descend to 1200m, expedite until passing 2400m, CES 6102.

4. CKK 9310,_____ FL256, expect descent at 20.

5. CES 2851, Wuhan Approach,_____ to WG and hold, maintain 1200m, inbound_____ 136 degrees, left hand_____, outbound time one minute, expect_____ clearance at 23.

6. B8913, Guanghan Approach,_____ visual between highway number 5 and iron bridge south of the airport.

7. CTL: SIA 253, descend when_____ to flight level, correction,_____ 9000ft, QNH 1005.

PIL:_____ FL 180 for_____, SIA 253.

Lesson 11-Exercise 3

III. Listening practice.

1. JAL 305,_____ to_____ via DE4 Arrival, descend to_____, report_____.

2. ACA 356,_____ to EX, descend and_____ 3000ft, report passing 5000ft.

3. BAW 403,_____ CH NDB above_____, expect NDB approach, runway_____, report CH NDB_____.

4. Maintain_____ flight level,_____ for descent at 45.

5._____ descent to FL 120 after 3 miles, report passing_____ levels.

6. DLH 720, continue descend to 1800m, expect join BX, hold as_____, maintain 1800m,_____ time 40.

7. PIL: Golf Arrival, ACA 321.

CTL:_____.

Unit Four
En-route

PIL：ACA 321, over TPF FL210, request descent.
CTL：_____.
PIL：Roger, maintain FL210 and standing by for descent.
8. PIL：Alpha Approach, AFR 626, descending to 9000ft, information M.
CTL：_____.
9. PIL：SIA 717, now reaching 7000ft.
CTL：_____.
PIL：Hold as published 7000ft, request holding instructions, SIA717.
CTL：_____.
PIL：Hold on LGS VOR/DME at 20 DME 7000ft, inbound track 210 degrees, turn left, outbound distance 24 miles DME.
CTL：That is correct.
10. PIL：Approach, AFR 288, good evening.
CTL：_____.
PIL：FL150, descending to FL100, information R received, AFR 288.
CTL：_____.
PIL：Descending to 6000ft, hold over MSK on reaching, AFR 288.

IV. Translation.
英译汉
1. Unable RVSM due equipment, resume normal separation.
2. ACA 505, Wuhan approach, cleared to hold over WG, maintain 1200m, inbound track 106 degrees, left hand pattern, outbound time 1 minute, expect further clearance at 23.
3. CSN 3572, descend to 4200m, expedite descent until passing 4800m, report passing even levels.
4. CCA 1578, hold at BCH, 2400m, inbound track 210, left hand pattern, outbound time one minute, expect approach time 12.
5. JAL 109, hold on the 170 radial of the NBW VOR between 25 and 35 miles DME, 7000ft, inbound track 350, right hand pattern, expect further clearance in 10 minutes.

汉译英
1. 国际1567,ADS-B机载设备工作好像不稳定/失效,终止ADS-B发送。
2. 国际101,在大王庄等待,高度五千四,入航航迹360,左航线,出航1分钟,预计进近时间15分。
3. 东方2289,北京进近,下降到修正海压幺八保持,修正海压幺洞幺两,保持目视间隔。
4. 国际101,在高度六千停止下降,由于高度五拐有活动,保持六千,预计进一步许可幺洞洞五。
5. 南方3392,在大王庄VOR一百八十度径向线距台35公里处等待,保持四千八,入航航迹三六洞,右航线,预计进一步许可洞八三五。

Unit Five

Approach

Lesson Twelve
RNAV, Approach

Lesson 12

Model Exchanges
A.
CTL: UAL 576, advise RNAV capability.
PIL: Negative/Affirm RNAV, UAL 576.
B.
PIL: Beijing Control, SAS 135, request RNAV.
CTL: SAS 135, RNAV approved/unable RNAV.
C.
PIL: ACA 927, unable RNAV due equipment.
CTL: ACA 927, roger, report able to resume RNAV.
D.
PIL: Juliett Approach, BAW 196, good morning.
CTL: BAW 196, Juliett Approach, go ahead.
PIL: BAW 196, reaching 3600m, information K received.
CTL: BAW 196, confirm reaching 3900m.
PIL: Negative, reaching 3600m, BAW 196.
CTL: BAW 196, squawk A3127, descend to 2400m, cleared to MEL, report reaching, reduce speed to 220 knots.
PIL: Squawking 3127, descending 2400m, reducing to 220 knots, cleared MEL, report reaching, BAW 196.
CTL: BAW 196, cleared ILS approach runway 13, no delay expected.
PIL: ILS approach, runway 13, BAW 196.

Unit Five
Approach

E.

PIL: Hong Kong Approach, CSN 3305, FL190, ETO TAMOT 1508, information D.

CTL: CSN 3305, Hong Kong Approach, cleared direct to Charlie FL190, enter controlled airspace FL190, hold at Charlie FL190 as published, expect approach time 35.

(Later)

CTL: CSN 3305, expect ILS approach runway 13.

PIL: ILS runway 13, CSN 3305.

CTL: CSN 3305, revised expected approach time 45.

PIL: Roger, 45, CSN 3305.

F.

PIL: Bravo Approach, CCA 1545, passing TR.

CTL: CCA 1545, Bravo Approach, expect visual approach runway 36R, direct to BS, descend and maintain 1800m, report runway in sight.

PIL: Visual approach runway 36R, maintain 1800m, CCA 1545.

PIL: CCA 1545, runway in sight.

CTL: CCA 1545, join right downwind runway 36R, you are number one in traffic, contact Tower on 118.1.

PIL: Right downwind runway 36R, 118.1, CCA 1545.

G.

PIL: Alpha Approach, JAL 101, over TX.

CTL: JAL 101, Alpha Approach, reduce speed to 260kts, report established (on the ILS).

PIL: Reducing to 260kts, JAL 101.

PIL: JAL 101, established ILS.

CTL: JAL 101, reduce to minimum approach speed, report outer marker.

PIL: Reducing to minimum, JAL 101.

PIL: Outer marker, JAL101.

CTL: JAL 101, contact Tower 118.7.

PIL: 118.7, JAL 101.

Dialogue

A.

PIL1: Wuhan Approach, CSN 3308, 4800m over ZF.

CTL: CSN 3308, Wuhan Approach, radar contact, turn left heading 180, descend and maintain 3900m.

PIL1: Heading 180, descending to 3900m, CSN 3308.

PIL2: Wuhan Approach, CES 5456, 5100m maintaining, passing HZ.

CTL: CES 5456, radar contact, reduce speed to 300kts.

Lesson Twelve
RNAV, Approach

PIL2: 300kts, CES 5456.

CTL: CSN 3308, maintain present altitude, reduce speed to 330kts.

PIL1: Roger, maintaining altitude, speed 330kts, CSN 3308.

CTL: CES 5456, descend to 3900m.

PIL2: Descending to 3900m, CES 5456.

CTL: CSN 3308, descend and maintain 2400 meters on QNH1011.

PIL1: Descending to 2400m on QNH1011, CSN 3308.

CTL: CES 5456, descend to 2100m on QNH1011.

PIL2: Descending to 2100m on QNH1011, CES 5456.

CTL: CES 5456, direct to DA and hold, maintain 2100m, give priority to CSN 3308 for landing.

PIL2: Roger, direct to DA and hold at 2100m, CES 5456.

CTL: CSN 3308, descend to 1200m, expect ILS approach, report established on the localizer.

PIL1: Roger, descending to 1200m, CSN 3308.

PIL1: Established on localizer, CSN 3308.

CTL: CSN 3308, reduce speed to 180kts, report outer marker, you are number 2.

PIL1: Reducing to 180kts, CSN 3308.

CTL: CSN 3308, reduce your speed to minimum, contact tower 118.7.

PIL1: Reducing to minimum, 118.7, CSN 3308.

CTL: CES 5456, fly heading 150, descend and maintain 1500m, radar vector for ILS approach runway 04.

PIL2: Roger, heading 150, descend and maintain 1500m, radar vector for ILS approach runway 04, CES 5456.

CTL: CES 5456, descend to 900m, report fully established.

PIL2: Descending to 900m, CES 5456.

PIL2: CES 5456, established.

CTL: CES 5456, continue approach, contact Tower 118.7, good day.

B.

PIL1: Wuhan Approach, CCA 1801, reaching 1800m, position 15km northeast of field, runway in sight, request visual approach runway 04.

CTL: CCA 1801, descend to 900m, cleared for visual approach runway 04.

PIL1: Roger, cleared for visual approach runway 04, descending to 900m, CCA 1801.

CTL: CSN 3108, maintain 2100m, direct to DA and join DA holding due traffic.

PIL2: Maintaining 2100m, direct to DA and join DA holding, CSN 3108.

PIL3: Wuhan Approach, SIA 204, 600m climbing.

CTL: SIA 204, Wuhan Approach, continue runway heading, climb to 1800m before setting

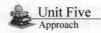

Unit Five Approach

heading to KG.

PIL3: Runway heading, 1800m before setting heading to KG, SIA 204.

PIL1: CCA 1801, reaching 900m.

PIL4: MAYDAY MAYDAY MAYDAY, ACA 235, starboard engine on fire, 12 miles southwest of your field, 900m descending, request landing priority.

CTL: ACA 235, Wuhan Approach, roger MAYDAY, cleared for straight-in ILS approach runway 04, you are number one.

PIL4: Straight-in ILS approach runway 04, ACA 235.

CTL: CCA 1801, Wuhan Approach, make one orbit right from your present position. Break Break, all stations, stop transmitting, MAYDAY.

New Words and Phrases

field *n.* 机场
revise *v.* 修正
proceed *v.* 飞往,行进
establish *v.* 建立
vector *v.* 雷达引导
make one orbit right 右转一圈
starboard engine 右引擎
port engine 左引擎
localizer back course 反航道

Notes

1. RNAV (Area Navigation): 该技术能使航空器在导航信号覆盖范围之内,沿任意期望的航迹飞行,不再受传统导航方法向、背导航台的飞行方法限制,航迹选择更加灵活,能有效地促进终端区飞行容量的增加,优化导航设施布局,提高运行水平。RNAV 导航允许航空器不飞经某些导航设施,它有以下三种基本应用:

(1) 在任何给定的起降点之间自主选择航线,以减少飞行距离,提高空间利用率;

(2) 航空器可在终端区范围内的各种期望的起降航径上飞行,以加速空中交通流量;

(3) 在某些机场允许航空器进行 RNAV 进近(如 GPS 进近落地),而无需那些机场的 ILS。

2. 进近航空器与进近管制单位建立初始联络时应报告本机位置和高度,管制员在收到航空器报告后通常向航空器指明进近方式。

3. 由于本场活动及其他原因不能立即确定进场航空器的进近方式时,管制员通常让航空器进入等待,然后讲明预计进近时间及进近方式。

4. downwind: 三边,起落航线的一部分; join right downwind: 加入右三边。

5. reduce speed to minimum: 减速到最小。调速是管制员调配冲突和间隔常用的管制

Lesson Twelve
RNAV, Approach

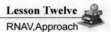

手段,驾驶员在收到指令后一般应复诵该指令。

6. 航空器进入管制区时必须进行雷达识别(radar identification),管制员才可发布后续指令。

Exercises

I. Substitution practice.

PIL: BeijingApproach, CSZ 8896, VYK 5700m, information D, request ILS approach.

CTL: CSZ 8896, cleared for ILS approach runway 36R, descend to 900m, QNH 1018, no delay expected.

— NDB
— VOR
— straight-in ILS
— visual
— localizer back course
— precision radar
— surveillance radar

PIL: Tango Approach, BAW 625, HGF 23, 4200m, Information D.

CTL: BAW 625, Tango Approach, radar contact, descend to 1800m on QNH1018, standard A2 arrival, expect ILS approach runway 17, number 3 in traffic, the latest information E.

— expect visual approach runway 35, direct to LM, descend and maintain 900m, QNH1012
— cleared VOR approach runway 09, descend to 900m on QNH1018, you are number one
— expect straight-in ILS approach runway 27, reduce speed to 210kts, QNH 1005
— direct to SEL and hold, 3300m, expect further clearance at 45

II. Fill in the blanks.

1. Cleared to MGA and descend to 2100m, expect ILS _____ runway 34.

2. Join A11 Arrival, descend and _____ 900m on QNH1001, no delay _____.

3. Expect visual approach runway 03, _____ to FD, descend and maintain 2700m, report runway in _____.

4. BAW 108, expect ILS _____ runway 09, you are number one, contact Tower 118.1 for _____ clearance.

5. Descend to FL128 before GHL, _____ until passing FL167.

6. CSN 3305, reduce your speed to _____, contact Tower 118.7.

7. SAS 956, cleared for ILS approach runway 06R, report established on _____.

8. ACA 751, 25 miles southwest of TON, FL138, _____ zone boundary 30, RTK 35, information G.

9. CCA 1301, cleared _____ approach runway 21, report _____.

10. SIA 917, stop descending immediately, I say again, stop descending, maintain present

Unit Five
Approach

_____. Break Break, DAL 304, descend now to 3000m on QNH 1013 and leaving FBT _____ 265.

III. Listening practice.

1. CSN 3108, cleared _____ runway 18L, _____ VMC, traffic Boeing 737 on 5km final.

2. CES 5402, _____ A02 Arrival, runway 36R, _____ 1800m.

3. AFR 188, after DA, cleared direct to _____, ILS approach runway 04, descend _____, you are number 1.

4. Chengdu Control, JAL 925, _____, TCAS RA.

5. BAW 751, cleared _____ ILS approach, runway 18, _____ to 1200m _____, report established on the localizer.

6. PIL: Alpha Approach, ACA541.
CTL: _____.
PIL: ACA 541, FL130, HRM 1235, information G.
CTL: _____.
PIL: Leaving FL130 for 7000ft QNH1015, ACA541.
PIL: ACA 541, HRM.
CTL: _____.
PIL: Cleared for ILS runway 17, ACA 541.

7. PIL: Foxtrot Approach, DLH 709.
CTL: _____.
PIL: DLH 709, estimating FHC VOR 1345 at FL108, request VOR approach runway 17 for training.
CTL: _____.
PIL: Roger, but is localizer back course available? DLH 709.
CTL: _____.
PIL: Cleared localizer back course, runway 17, DLH 709.

8. PIL: Charlie Approach, JAL 414, over TEL, maintaining 2100m.
CTL: _____.
PIL: NDB runway 32, leaving 2100m for 900m, JAL 414.
PIL: JAL 414 now MB, airport in sight, request visual approach.
CTL: _____.
PIL: Continue NDB approach, JAL 414.

IV. Translation.

英译汉

1. CCA 1579, expect visual approach runway 36R, direct to QU, descend and maintain 1800m, report runway in sight.

Lesson Twelve
RNAV,Approach

2. CES 5321, continue approach, I will keep you advised.

3. CSN 3308, Wuhan Approach, roger MAYDAY, the nearest suitable airport is Wuhan Tianhe airport, cleared direct to DA, descend and maintain 2700m, we will inform Wuhan Tower to get ready for your landing.

4. DAL 901, cleared VOR approach runway 18R, descend at your own discretion, you are number one, report runway in sight.

5. CES 5402, Wuhan approach, cleared to WG and hold as published, expect further clearance at 20.

汉译英

1. 二发强烈抖动,原因不明,我们随时可能关闭发动机,申请立即进近。

2. 东方5301,武汉进近,沿R10D RNAV 离场。

3. 南方3342,广汉进近,可以直线NDB进近,跑道两两。

4. 在RAL VOR 300度径向线距台25到30海里间等待,高度三千,入航航迹120,左航线,延误时间不定。

5. 东方5301,武汉区域,不能实施RNAV运行,雷达引导直飞武汉VOR。

Lesson Thirteen
TCAS, Traffic Circuit

Model Exchanges
A.

当 TCAS 提示有冲突时

PIL：Beijing Control, CSN 3579, TCAS climb.

CTL：CSN 3579, Beijing Control, roger, continue climb and maintain 11000m.

B.

当 TCAS 冲突消除后

PIL：Beijing Control, CSN 3579, clear of conflict, returning to 9500m.

CTL：CSN 3579, roger, continue descend and maintain 8400m.

C.

当 TCAS 提示不能执行管制指令时

CTL：JAL 925, Beijing Control, descend to 7200m.

PIL：Beijing Control, JAL 925, unable, TCAS RA.

D.

PIL：Beijing Approach, CCA 1572, turning final.

CTL：CCA 1572, continue approach, report short final, you are number 2, a B737 crossing the threshold, contact Tower on 118.5.

PIL：Roger, 118.5, CCA 1572.

E.

PIL：Nanjing Approach, BAW 891, downwind.

CTL：BAW 891, Nanjing Approach, make long approach, you are number two, number one A320 5km on final, report traffic in sight.

PIL:Traffic in sight,BAW 891.

CTL:BAW 891,follow the A320,cleared for visual approach runway 06.

F.

CTL:G-ABCD,make one orbit right due traffic on the runway,report again on final.

PIL:G-ABCD,orbiting right.

PIL:G-ABCD,final.

CTL:G-ABCD,continue approach,wind 180 degrees 5 knots.

G.

PIL:Beijing Tower,CCA 982,1500m for landing.

CTL:CCA 982,make straight-in ILS approach runway 36L.

PIL:Straight-in runway 36L,CCA 982.

CTL:CCA 982,wind 330 degrees 8m/s,QNH 1011,cleared to land.

Dialogue

A.

PIL:Wuhan Approach,CSN 3506.

CTL:CSN 3506,go ahead.

PIL:Wuhan Approach,CSN 3506,15km south of your field heading 020,1500m on QNH 1016,ETA Wuhan 25 for landing.

CTL:CSN 3506,cleared to Wuhan,VFR to join right downwind,runway 18R,wind 170 degrees,10m/s,QNH 1016,report joining downwind.

(Later)

PIL:Wuhan Approach,CSN 3506,downwind.

CTL:CSN 3506,make long approach,you are number two,number one B777 4km final,report traffic in sight.

PIL:Number two,traffic in sight,CSN 3506.

CTL:CSN 3506,follow number one,make visual approach runway 18R.

PIL:Roger,visual approach 18R,CSN 3506.

CTL:CSN 3506,maneuver right and left for spacing,you are too close to preceding traffic.

PIL:Maneuvering right and left,CSN 3506.

PIL:CSN 3506,final.

CTL:CSN 3506,wind 160 degrees 8m/s,cleared to land.

PIL:Cleared to land,CSN 3506.

B.

PIL:Wuhan Approach,CES 3218,request radar vector to Wuhan.

CTL:CES 3218,for identification turn left heading 210.

PIL:Left heading 210,CES 3218.

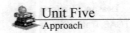

Unit Five
Approach

CTL:CES 3218, radar contact, 20km northwest of the field, turn left heading 180 for downwind.

PIL:Left heading 180, CES 3218.

CTL:CES 3218, met report of Wuhan: wind 340 degrees, 8 knots, visibility 8km, broken 600m, QNH 1010.

PIL:QNH1010, CES 3218.

CTL:CES 3218, vectoring for left circuit, visual approach, RWY 36.

PIL:Left circuit, visual approach, RWY 36, CES 3218.

CTL:CES 3218, turn left heading 090 for base.

PIL:Left heading 090, CES 3218.

CTL:CES 3218, position base, 25km southwest of the field.

PIL:CES 3218.

CTL:CES 3218, turn left heading 030.

PIL:Left heading 030, CES 3218.

CTL:CES 3218, turn left heading 360 for final.

PIL:Left heading 360, CES 3218.

CTL:CES 3218, position final, 20km from touchdown.

PIL:CES 3218.

CTL:CES 3218, descend to 600m, report runway in sight.

PIL:Runway in sight, CES 3218.

CTL:CES 3218, position 15km from touchdown, cleared for visual approach, RWY 36.

PIL:Cleared for visual approach, RWY 36, CES 3218.

CTL:CES 3218, contact Wuhan Tower 130.0, good day.

PIL:Wuhan Tower 130.0, CES 3218.

New Words and Phrases

traffic circuit (pattern) 起落航线

upwind *n.* 一边

crosswind *n.* 二边

downwind *n.* 三边

base *n.* 四边

final *n.* 五边

turning base 三转弯

orbit *n. & v.* 圈,圆圈,环形;盘旋

spacing *n.* 腾……空间

preceding *adj.* 位于……之前,前面的

short approach 小航线进近

long approach 大航线进近
turning final 四转弯
maneuver v. 机动(操作)

Notes
1. 标准起落航线是左起落航线。如果起落航线方向有变化时需要指明。
2. 如果当时条件允许,管制员可以让飞机做直线进近。
3. 为了调配起落航线上飞行冲突,管制员会发布延缓或加速指令:
（1）延缓通常是延长三边;
（2）延缓的另一种方式是在当前位置左转或右转一圈;
（3）当两机在五边的安全距离不够时,让后边的飞机做左右机动飞行;
（4）加速的一般做法是让飞机作小航线进近,即尽量缩短三边,及早加入五边。
4. 起落航线,示意图见下图
upwind:一边,离场边,起飞;
crosswind:二边,侧风边,此时飞行航迹线与跑道垂直并大体与风向交叉,90°转弯,出航;
downwind:三边,下风边,与跑道平行,并沿着与着陆航向相反的方向飞行,顺风飞行,巡航;
base:四边,基线边,降落的过渡部分,开始下降高度,进场;
final:五边;
long final(长五边):当起落航线上的航空器在距接地点达7公里（4海里）以前转到五边上;或当直接进近的航空器距接地点15公里(8海里) 时,报告"长五边"。在这两种情况下,航空器应距接地点7公里时,报告"五边"(final)。

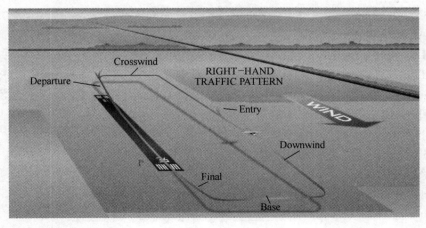

Exercises
I. Fill in the blanks.
1. CSN 3392, Luoyang Tower, number 2, _____ B757 on final.

Unit Five
Approach

2. CSN 3308, Wuhan Tower, make _____ approach, runway 22, wind 210, 12m/s, temperature 32, QNH 1010.

3. CTL: CSN 3342, _____ downwind, you are number two, follow B747 8km on _____ .

PIL: _____ , number two, traffic in _____ , CSN 3342.

4. Make a right _____ and _____ downwind.

5. CSN 304, _____ right and left for _____ , you are too close to _____ traffic.

6. CSN 308, _____ right downwind runway 15, wind 170 10knots, QNH1001.

7. ACA 230, joining downwind, _____ in sight, request _____ approach runway 35.

8. G-CD, descend to _____ height 1000 feet, QNH1006, join right downwind, runway 23, report _____ .

9. CSN 3167, make one _____ right, report again on _____ .

10. JAL 820, Wuhan Tower, _____ approach, report _____ .

Ⅱ. **Listening practice.**

1. PIL: Beijing Approach, ACA 290, _____ final.
CTL: ACA 290, _____ to land, _____ 240 10knots.

2. CTL: G-CD, number one, make _____ .
PIL: Short approach, _____ .

3. CSN 3460, _____ DC-9 _____ and ATR-42 downwind.

4. CCA189, change to runway 36L due _____ unable to _____ runway immediately, report _____ .

5. Wuhan Tower, CSN 3802, _____ south of your field for _____ .

6. CSN 3107, _____ of runway _____ . _____ , BAW 421, make _____ at your present position, standby for _____ information.

7. CSN 3732, join _____ runway 31, wind _____ , QNH 1010, report turning final.

Lesson 13-
Exercise 2

104

8. PIL: CSN 3561, downwind.

CTL: _____ .

PIL: Number two, traffic in sight, CSN 3561.

PIL: _____ .

9. PIL: Beijing Tower, CES 2910, long final

CTL: _____ .

CTL: CES 2910, go around, standard procedure, there's a runway lighting failure.

PIL: _____ .

10. PIL: Tower, JAL 556, 15 miles south of field, 1200m, information L for landing.

CTL: _____ .

PIL: Roger, down to 600m on QNH1012, right hand pattern, runway 25, JAL 556.

PIL: JAL 556, downwind.

CTL: _____ .

PIL: Traffic in sight, JAL 556

PIL: _____ .

CTL: JAL 556, report final.

PIL: JAL 556, final.

CTL: _____ .

Ⅲ. Translation.

英译汉

1. CSN 3301, number 2 to land, number 1 at touchdown, report outer marker.

2. ACA 185, turn left from your present position, circle to land runway 04, you are number one.

3. CCA 1834, join right downwind runway 15, wind 170 degrees, 10 knots, QNH1001.

4. AFR 501, maneuver right and left for spacing, you are too close to preceding traffic.

5. KAL 889, descend to circuit height 900m, QFE 998.

汉译英

1. 东方5230,延长三边,你是第二个落地,跟在五边5公里的B737后面。

2. 英航857,右转一圈,因为冲突,跑道上有飞机,重新回到五边报告。

3. 福州塔台,我们无法在决断高对准跑道中线,可能是由于阵风,现在一切都正常了,南方3245。

4. 日航123,延长三边两公里,给离场航空器腾空间,三转弯报告。

5. 南方3385,塔台,按照标准程序复飞,稍等进一步指令。

Unit Six

Landing and After Landing

Lesson Fourteen
Missed Approach, Diversion and Local Training

Lesson 14

Model Exchanges

A.

PIL: AFR 170, outer marker.

CLT: AFR 170, wind 140, 7 knots, cleared to land.

PIL: Cleared to land, AFR 170.

CLT: AFR 170, go around, the runway is still blocked, previous landing A321 unable to vacate runway.

PIL: Going around, AFR 170.

B.

PIL: ANA 793, final.

CTL: ANA 793, continue approach, report outer marker.

CTL: ANA 793, advise go around, crosswind 23kts, gusting 30kts, preceding CSZ 8516 has diverted to its alternate.

PIL: No problem, we can make it.

CTL: ANA 793, runway 36L, cleared to land.

C.

CTL: ACA 108, Beijing Capital International Airport closed for runway lighting system failure, hold at VYK, EAT 2200.

PIL: ACA 108, roger.

PIL: Beijing Approach, ACA 108, holding at VYK, unable to hold for more than 45 minutes, request divert to Tianjin.

CLT: ACA 108, standby for further clearance.

109

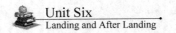
Unit Six
Landing and After Landing

PIL: Standing by, ACA 108.

D.

PIL: CES 5101, short final.

CTL: CES 5101, runway 24, RVR 650 meters, cleared to land.

PIL: Cleared to land, CES 5101.

PIL: No contact at minima, going around, CES 5101.

CTL: CES 5101, climb straight ahead to 600m on QNH, contact Approach on 125.8.

PIL: Straight ahead to 600m, 125.8, CES 5101.

E.

PIL: CCA 1787, nose wheel appears jammed, request touch and go to jar the wheel down and visual gear inspection.

CTL: CCA 1787, cleared touch and go.

PIL: Cleared touch and go, CCA 1787.

CTL: CCA 1787, nose gear does not appear down, direct to the holding area to have a complete check.

PIL: Wilco, CCA 1787.

F.

PIL: CDG 6782, request touch and go to make another approach training.

CTL: CDG 6782, negative due weather, a thunderstorm is approaching the airport, make full stop, runway 19, cleared to land.

PIL: Full stop landing runway 19, CDG 6782.

Dialogue

A.

PIL1: PANPAN PANPAN PANPAN, Wuhan Tower, CSN 3501, one passenger is very seriously ill, position 25km south of the airport, 1500m descending, request priority landing.

CTL: CSN 3501, Wuhan Tower, roger PANPAN, ILS approach runway 04 at your convenience, report outer marker.

PIL1: Runway 04, will call you outer marker, CSN 3501.

PIL2: Wuhan Tower, CSN 3342, approaching holding point runway 04, request line up.

CTL: CSN 3342, hold short of runway, PANPAN in progress.

PIL2: Holding short, CSN 3342.

CTL: AFR 188, extend downwind one minute to give room for the emergency traffic.

PIL3: Roger, extend downwind one minute, AFR 188.

PIL4: Wuhan Tower, SAS 996, we are going around, we encountered a sudden loss of height just now, possibly windshear.

CTL: SAS 996, follow standard missed approach procedure.

Lesson Fourteen
Missed Approach, Diversion and Local Training

PIL1: Wuhan Tower, CSN 3501, outer marker.

CTL: CSN 3501, caution, preceding aircraft reported windshear along your landing course, wind variable, runway 04, cleared to land.

PIL1: Cleared to land, CSN 3501.

B.

PIL1: Tower, CSN 3289, heavy, 11km south at 1500m, information G for landing.

CTL: CSN 3289, descend to circuit altitude 900m on QNH 1012, join left downwind runway 21, report joining.

PIL1: Roger, down to 900m on QNH 1012, left-hand pattern runway 21, CSN 3289.

PIL1: CSN 3289, downwind.

CTL: CSN 3289, report base, number 2, Airbus on the 5 miles final.

PIL1: Traffic in sight, CSN 3289.

PIL2: Tower, CQH 9510, runway in sight, request touch and go for training.

CTL: CQH 9510, negative for touch and go due flow control, we are waiting for some special flights, make full stop, cleared to land.

PIL2: Roger, cleared for full stop landing, CQH 9510.

PIL1: Turning base, CSN 3289.

CTL: CSN 3289, report final.

PIL1: CSN 3289.

PIL1: Final, CSN 3289,

CTL: CSN 3289, continue approach, wind 230, 10 knots.

New Words and Phrases

go around 复飞

missed approach 复飞

diversion *n.* 改航,备降

short final 短五边

long final 长五边

contact *n.* 观察,看见

minima *n.* 最低限度,最小值

touch and go 落地连续

full stop 全停

priority *n.* 优先

nose gear appears up/down 前轮看上去收上/放下

Notes

1. 复飞 go around 属于指令性管制指挥用语,仅用于发布复飞指令时使用,其他语句

Unit Six
Landing and After Landing

中应使用 missed approach。

go around (most commonly used)

pull up (used in emergency)

overshoot (used in the military)

missed approach (used in missed approach phase or missed approach procedure)

2. 在下列情况下应建议机组复飞：

Poor visibility

Blocked runway

Bad approach (altitude too high or low /wheels not down)

Too close to preceding aircraft

Strong cross wind

Bad weather (heavy rain, snow, hailstorm, etc.)

Runway lighting failure

No landing clearance was received (in a controlled field)

…

3. 复飞指令含强制性指令和建议性指令。

(1) 强制性复飞指令是指管制员发现目前情况对飞行安全构成威胁而向机组发出的复飞命令。机组必须执行，通常管制员会给出复飞原因。

(2) 当管制员观察到飞机飞行情况不正常或有其他不安全因素时，会向机组发出建议性复飞指令，着陆或者复飞由航空器驾驶员最后决定，并且对其决定负责。

4. no contact at minima：最低高度（决断高度）无法能见地面。contact 在这里当"看见，观察，能见"讲。

Exercises

I. Substitution practice.

1. CTL: CSN 3105, clear to land runway 15.

PIL: No contact at minima, <u>going around</u>.

— overshooting

— pulling up

— executing missed approach

2. CTL: Go around, I say again, go around, <u>aircraft on the runway</u>.

— the runway is blocked

— too close to the preceding aircraft

— wheels are not down

— the preceding aircraft has not vacated runway

3. PIL: BAW 108, holding at VYK, <u>unable to hold for more than 45 minutes</u>, request diversion to Hohhot.

Lesson Fourteen
Missed Approach, Diversion and Local Training

— thunderstorm is on the field
— technical problems with the aircraft

II. Fill in the blanks.

1. CSN 1509, _____ due vehicle broken down on the runway.

2. We are going around, we encountered a sudden _____ of height just now, possibly _____ .

3. Dalian aerodrome _____ for snow clearance, hold at MK.

4. CSN 3702, unable to _____ for more than 1 hour, request _____ to Tianjin.

5. PIL: CSN 3108, _____ marker.
CTL: CSN 3108, _____ go around, altitude too high.
PIL: No problem, we can _____ it.
CTL: CSN 3108, _____ to land runway 12.

6. CSN 3105, cleared _____ to make another _____, report downwind.

7. CTL: CES 2503, you are number one to _____ . Caution, windshear reported at 300m 4km _____, runway 07.
PIL: Number one to land, CES 2503.
(Later)
PIL: _____, CES 2503.
CTL: CES 2503, standard _____, when passing 600m, turn right to XLN VOR.

III. Listening practice.

1. No contact at minima, _____ .

2. CSN 3601, make _____ due to traffic _____, cleared to land wind _____ .

Lesson 14-
Exercise 3

3. PIL: Chengdu Approach, CSN 3509, holding, request _____ .
CTL: CSN 3509, your EAT will be _____, standby.
PIL: Chengdu Approach, CSN 3509, _____ to hold for more than _____, request _____ to Guiyang.

4. Request _____ approach for another approach _____ .

5. We can't _____ the nose gear, going around.

6. PIL: DLT, turning _____ .
CTL: DLT, _____ .
PIL: Roger, DLT
CTL: DLT, go around, the B737 is unable to _____ the runway.
PIL: _____, DLT.
CTL: DLT, _____ to 1200m, contact _____ .

Unit Six
Landing and After Landing

IV. Translation.

英译汉

1. CES 2827, runway is blocked, go around immediately, I say again, go around.

2. CSN 3210, negative for touch and go due to traffic congestion, make full stop, runway 21, cleared to land.

3. ACA 108, Shenyang Approach, aerodrome closed due snow clearance in progress, cleared to divert to Changchun.

4. CCA 1509, we encountered severe CAT, some passengers have been injured, request divert to Meilan airport which is closest to us.

5. Tower, B2890, runway in sight, request touch and go for training.

汉译英

1. 武汉塔台，日航223，我们复飞了，由于液压系统故障，我们无法放下起落架。
2. 北京塔台，英航411，由于鸟吸左发失效，我们返航了，申请优先着陆。
3. 厦门塔台，白鹭8837，最低高度无法能见地面，复飞了。
4. 南京塔台，海南7374，我们无法摇动放下起落架，我们打算复飞，直飞等待区检查。
5. 香港塔台，日航520，申请低空通场目视检查起落架。

Lesson Fifteen
Landing and After Landing, Low Altitude Warning, Terrain Alert

Lesson 15

Model Exchanges
A.
PIL:AFR 271,final.
CTL:AFR 271,continue approach.
PIL:Continue approach,AFR 271.
PIL:AFR 271,outer marker.
CTL:AFR 271,wind 150,7m/s,cleared to land.
B.
PIL:Hongqiao Tower,CSN 3781,ILS approach,runway 36L.
CTL:CSN 3781,continue approach,B757 just rolling.
PIL:Continue approach,CSN 3781.
CTL:CSN 3781,cleared to land.
PIL:Cleared to land,CSN 3781.
CTL:CSN 3781,take first right and contact Ground 119.2 when vacated.
C.
CTL:CSC 8223,take second left and contact Ground on 121.5.
PIL:Take the second left,121.5,CSC 8223.
　　PIL:Nanjing Ground,CSC 8223,good morning,we seem to have a nose gear tyre blow out on landing,request a tug to tow us to the apron.
　　CTL:CSC 8223,can you move forward under your own power until you're past the next intersection?
　　PIL:Affirm,I think we can manage that,CSC 8223.

115

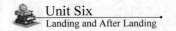
Unit Six
Landing and After Landing

CTL: CSC 8223, we'll get a tug out to you as soon as possible.

D.

CTL: BAW 591, backtrack the runway, exit 5 to vacate the runway, contact Ground on 118.4.

PIL: Backtrack the runway, exit 5, 118.4, BAW 591.

PIL: BAW 591, runway vacated.

CLT: BAW 591, taxi to Gate B6 via taxiway C1, B2 and D2.

PIL: C1, B2 and D2 to Gate B6, BAW 591.

E.

CTL: JAL 684, Alpha Ground, taxi with caution, taxiway E2 is partly covered with slush and ice, braking action poor, an arriving EMB-145 just shot off the taxiway.

PIL: JAL 684, roger, thanks for the alert.

F.

PIL: Wuhan Tower, CAL 218, catering has not finished yet, besides we have one passenger missing, we'll have to unload his checked baggage, please delay our slot time until 1530.

CTL: CAL 218, roger.

G.

CTL: CSN 3392, follow green light.

PIL: Follow green light, CSN 3392.

H.

CTL: CSN 3392, cancel green light guidance, follow voice instructions, continue taxi via E, hold short of Y7.

I.

CTL: CSN 3392, stop-bar unserviceable, cross red stop-bar, via G1 line up and wait, runway 35R.

PIL: Cross red stop-bar, via G1 line up and wait, runway 35R, CSN 3392.

J.

CTL: CXA 9123, low altitude warning, check your altitude immediately, QNH is 1008, the minimum safety altitude is 600m.

PIL: Wilco, thanks for the alert, CXA 9123.

K.

PIL: Beijing Approach, CCA 101, terrain alert, pulling up, climbing to 1200m, QHN 1001.

CTL: CCA101, continue climb and maintain 1800m.

Dialogue

A.

PIL1: Beijing Tower, CSN 3106, outer marker.

Lesson Fifteen
Landing and After Landing, Low Altitude Warning, Terrain Alert

CTL: CSN 3106, wind 320 at 9m/s, runway 36R, QNH 1013, cleared to land.

PIL1: Landing runway 36R, 1013, CSN 3106.

PIL2: Beijing Tower, BAW 411, approaching holding point runway 36R.

CTL: BAW 411, hold short of runway.

PIL2: Holding short, BAW 411.

PIL2: Beijing Tower, BAW411, ready for departure.

PIL3: MAYDAY MAYDAY MAYDAY, Beijing Approach, JAL 123, we have been hit by an unknown object and the side cabin window was broken, two passengers were seriously injured, 16 kilometers north of your field, request priority landing and emergency services

CTL: JAL123, Beijing Approach, roger MAYDAY, cleared straight-in ILS approach runway 36R, you are number one, Break Break, all stations, stop transmitting, MAYDAY.

CTL: BAW 411, standby, I will call you back.

PIL2: Standing by, BAW411.

(Later)

CTL: BAW 411, are you ready for immediate departure?

PIL2: Affirm, BAW 411.

CTL: BAW 411, after departure, climb on runway heading to 300m, then turn right to HG at 900m, wind 320 at 9m/s, cleared for take-off.

PIL2: Runway heading 300m and turn right to HG at 900m, cleared for take-off, BAW 411.

B.

PIL: Pudong Tower, JAL 901 is fully established runway 35.

CTL: JAL 901, report outer marker.

PIL: JAL 901.

(Later)

PIL: Outer maker, JAL 901.

CLT: JAL 901, continue approach, B737 just rolling.

PIL: JAL 901.

CTL: JAL 901, wind 330 8m/s, cleared to land.

PIL: Cleared to land, JAL 901.

CTL: JAL 901, landing time 1405, backtrack the runway, take first right.

PIL: Backtracking, first right, JAL 901.

CTL: JAL 901, expedite vacating runway, aircraft on short final.

PIL: Expediting, JAL 901.

CTL: JAL 901, contact Ground 120.3 when vacated.

PIL: 120.3, JAL 901.

Unit Six
Landing and After Landing

New Words and Phrases

vacate runway 脱离跑道
fully *adv.* 完全地
unknown object 不明物体
marshaller *n.* 信号员
low altitude warning 低高度告警
terrain alert 近地告警
blow out(轮胎)爆破
cabin *n.* 客舱

Notes

1. just rolling 是 just rolling for take-off 的省略讲法,意思是"正在起飞滑跑"。

2. acknowledge by flashing landing lights:闪烁着陆灯表示收到。

3. is fully established:The aircraft is in line with the localizer beam and on the glide path. 飞机对准了航向道和下滑道。

4. Factors affecting the landing of aircraft:
   (1) Wind
   (2) RWY conditions
   (3) Weight of aircraft
   (4) Temperature
   (5) QNH
   …

5. Ground Proximity Warning System (GPWS) 近地告警系统:A system carried on many aircraft to warn the pilot that the aircraft may be in danger of inadvertent contact with the ground. It is intended to reduce the occurrence of controlled-flight-into-terrain (CFIT) accidents,in which aircraft with no apparent mechanical difficulty or defect strike the ground while under the direct or indirect control of the pilot. These accidents usually occur in conditions of poor visibility due to atmospheric obscuration such as fog or rain,or darkness of night.

6. "follow green light" 意为"跟随绿灯滑行",是 A-SMGCS(高级场面活动引导控制系统)用语。

Exercises

Ⅰ. Fill in the blanks.

1. JAL 120,leave CH _____ 320 for Lake,_____ approach 125.8.

2. CSN 3679,Beijing Approach,_____ 2100m,radar _____ ILS approach runway 16,QNH 1010.

3. CES 2890,turn left _____ 210 to _____ ILS runway 16, report _____ on

Lesson Fifteen
Landing and After Landing, Low Altitude Warning, Terrain Alert

the localizer

4. CSN 3380, _____ first left and _____ Ground 121.6 when _____ .
5. CSN 3901, landing _____ 1208, _____ the runway, take first right.
6. PIL: Beijing Tower, CSN 3982, on _____ .
CTL: CSN 3982, _____ wind 310 degrees 7m/s, _____ to land runway 36L.
7. PIL: GCB, _____ marker.
CTL: GCB, continue _____ A320 ready for departure.
PIL: GCB.
CTL: GCB, _____ land, wind calm.
PIL: Cleared to land, GCB.
CTL: GCB, landing time 1212, _____ first right and contact ground 121.0 when _____ .

II. Listening practice.

1. AFR 677, descend on the _____ , maintain not less than _____ to outer marker, contact Tower _____ .
2. CSN 3901, taxi on _____ until reaching _____ , then turn left proceed to _____ , report _____ .
3. Your position is _____ from touchdown, continue approach on the _____ to the outer marker, contact Tower on _____ .
4. PIL: Nanjing _____ , CSN 3870, runway _____ .
CTL: CSN 3870, turn _____ onto taxiway _____ , proceed to _____ , parking _____ .
PIL: _____ B4 to P2, stand 30, CSN 3870.
5. CTL: ACA 701, take the _____ and contact Ground on _____ .
PIL: 121.5, ACA 701.
PIL: Wuhan Ground, ACA 701, _____ .
CTL: ACA 701, take the _____ inner taxiway, to stand _____ .
PIL: Second left, inner taxiway, stand Delta 7, ACA 701.
6. CTL: CSN 3251, _____ .
PIL: Cleared to land, CSN3251.
(*pause*)
CTL: Shenzhen Tower, CSN 3251, we aquaplaned after touch-down and have at least 2 tyres blown out on right main gear, _____ , please advise company maintenance _____ .
7. PIL: Outer marker, AFR 901.
CTL: _____ .

Lesson 15-Exercise 2

119

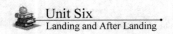

PIL: Cleared to land, AFR 901
CTL: _____.
PIL: Going around, AFR 901
CTL: _____.
PIL: 1200m for right downwind runway 15, AFR 901.

Ⅲ. Translation.
英译汉

1. Beijing Ground, AIC 182, we have just skidded off taxiway D, left main gear appears to be stuck in the mud, can you send a tug around?

2. AFR 671, take second right and contact Ground 119.2 when vacated.

3. Beijing Tower, KLM 057, we are not familiar with the local airport, request a FOLLOW-ME car to guide us to the parking stand.

4. CSN 3109, caution, cross wind 280 degrees 8 knots, cleared to land.

Guiyang Approach, OKA 8509, terrain alert, pulling up, climbing to 1500m on QNH 1013.

汉译英

1. 日航850，落地时间四五分，在跑道两幺上调头。

2. 东方5820，下一个道口右转脱离跑道，沿滑行道B4、D1和C1滑到停机位洞四。

3. 国际1718，注意！前机报告在你着陆航径上有风切变，跑道18左，可以落地。

4. 北京塔台，东方2837，申请无起落架着陆，请在跑道上铺设泡沫毯。

5. 汉莎917，西安塔台，我们已通知医疗部门(急救部门)，医生和救护车将在客机坪等候你们。

Unit Seven

Radar Control

Lesson Sixteen
Radar Identification and Vectoring, Precision Radar Approach

Model Exchanges Lesson 16

A.

CTL: CSZ 7981, report your heading and altitude.

PIL: Heading 185 at 3900m, CSZ 7981.

CTL: CSZ 7981, for identification turn right heading 210.

PIL: Right heading 210, CSZ 7981.

CTL: CSZ 7981, identified, 15km northwest of VY, continue present heading.

PIL: CSZ 7981.

B.

CTL: KAL 630, report your heading.

PIL: Heading 190, KAL 630.

CTL: KAL 630, roger, turn left heading 080 for separation.

PIL: Leftheading 080, KAL 630.

CTL: KAL 630, make a three sixty turn left for delaying action.

PIL: Three sixty turn left, KAL 630.

C.

CTL: CQH 9305, squawk 6203.

PIL: Squawking 6203, CQH 9305.

CTL: CQH 9305, confirm squawk.

PIL: Squawking 6203, CQH 9305.

CTL: CQH 9305, identified, continue present heading, omit position report until VYK.

PIL: CQH 9305.

Unit Seven
Radar Control

D.

CTL: CAL 918, report your heading and flight level.

PIL: CAL 918, heading 135 at FL330.

CTL: CAL 918, identified, fly heading 190.

PIL: Heading 190, CAL 918.

CTL: CAL 918, radar service terminated, resume own navigation, direct to CH.

PIL: Direct CH, CAL 918.

E.

CTL: CCA 101, radar contact, abeam CH, continue heading 310.

F.

CTL: CES 2289, not identified, continue own navigation, direct to CH.

G.

CTL: CCA101, comply with speed restrictions, do not exceed Mach point eight five.

H.

CTL: CCA 101, radio contact lost at 25, if you read, turn right heading 250.

CTL: CCA 101, turn observed, position 20km south of VYK, will continue radar control.

I.

CTL: CES 2289, Nanjing Approach, unable to spot SSR response, advise transponder capability.

PIL: Nanjing Approach, CES 2289, transponder is Mode Charlie.

J.

CTL: CSN 3392, this will be a precision radar approach, runway 27, obstacle clearance altitude 400ft, position 6 miles east of Georgetown, turn right heading 260, descend to 2500ft, QNH 1012.

K.

CTL: JAL 505, Hong Kong Precision, how do you read?

PIL: Read you 5, loud and clear, JAL 505.

CTL: JAL 505, do not acknowledge further transmissions.

Approaching glide path, heading is good.

Check your minima.

On glide path, 5 miles from touchdown.

…

Dialogue

A.

CTL: CES 2539, report heading and level.

PIL: Heading 134, FL160, CES 2539.

Lesson Sixteen
Radar Identification and Vectoring, Precision Radar Approach

CTL: CES 2539, for identification, turn right heading 165.

PIL: Right heading 165, CES 2539.

CTL: CES 2539, identified, 10 miles northwest of CH, fly heading 190.

PIL: Heading 190, CES 2539.

CTL: CES 2539, confirm your heading.

PIL: Heading 190, CES 2539.

CTL: CES 2539, continue heading 190 to cross Mike.

PIL: Wilco, CES 2539.

CTL: CES 2539, turn left heading 105.

PIL: Left 105, CES 2539.

CTL: CES 2539, hold at Charlie, radar contact lost due radar failure, repair is under way, remain this frequency.

PIL: CES 2539.

CTL: CES 2539, radar resuming operation, fly heading 030, descend to 6000 feet on QNH 1012.

PIL: Heading 030 and descend to 6000 feet on QNH 1012, CES 2539.

CTL: CES 2539, make a left orbit for sequencing.

PIL: Orbiting left, CES 2539.

CTL: CES 2539, resume own navigation, direct CH track 030 15 miles.

PIL: 030 15 miles direct CH, CES 2539.

B.

PIL: HongKong Approach, CSN 3306, FL150, TAMOT 15, information B.

CTL: CSN 3306, Hong Kong Approach, squawk ident.

PIL: Squawking, CSN 3306.

CTL: CSN 3306, now turn right heading 190.

PIL: Right heading 190, CSN 3306.

CTL: CSN 3306, report speed.

PIL: Speed 220kts, CSN 3306.

CTL: CSN 3306, reduce speed to 190kts.

PIL: Reducing to 190kts, CSN 3306.

CTL: CSN 3306, turn left heading 090 to CH.

PIL: Left heading 090 to CH, CSN 3306.

CTL: CSN 3306, traffic, 2 o'clock, 6 miles, westbound, fast moving.

PIL: Looking, CSN 3306.

PIL: Traffic in sight, it looks like a helicopter, now passed clear, CSN 3306.

CTL: CSN 3306, turn left heading 030 and descend to 6000 feet, you are No. 3.

PIL: Heading 030, descending to 6000 feet, CSN 3306.

Unit Seven
Radar Control

CTL:CSN 3306,turn right heading 045,cleared IGS runway 13,report established.
PIL:Heading 045,IGS runway 13,CSN 3306.
PIL:IGS established,CSN 3306.
CTL:CSN 3306,14 miles from touchdown,contact Tower on 118.7.
PIL:118.7 for Tower,CSN 3306.

New Words and Phrases

identify *v.* 识别
identification *n.* 识别
radar identification 雷达识别
identified *adj.* 雷达已识别
radar contact 雷达看到
vector *v.* 引导
resume *v.* 恢复
navigation *n.* 领航,导航,航行
omit *v.* 省略
delaying action 推迟进近的机动飞行
squawk *v.* 调置应答机
temporarily *adv.* 暂时地
remain *v.* 保持
remain this frequency 当前频率保持长守
sequence *n.* 次序 *v.* 排序
track *v. & n.* 出航,航迹
leg *n.* 航线之一段
bound *adj.* 向……方向,往……去的
westbound *adj.* 向西
RPI(Radar Position Indication) 雷达位置指示
RPS(Radar Position Symbol) 雷达位置符号
PSR blip 一次监视雷达回波
SSR response 二次监视雷达应管
PSR(Primary Surveillance Radar) 一次监视雷达
PAR(Precision Approach Radar) 精密进近雷达
SSR(Secondary Surveillance Radar) 二次监视雷达
ASR(Airport Surveillance Radar) 场面监视雷达

Notes

1. radar contact 是 identified 的另一说法,国际民航组织推荐使用后者。

126

Lesson Sixteen
Radar Identification and Vectoring, Precision Radar Approach

2. 雷达引导结束时管制员通知驾驶员恢复自主领航,同时视情况通报飞机位置并发布相关指令。

3. 管制员有时指示飞机做360°转弯,借此达到化解冲突或调序的管制目的。

4. Precision radar approach (PAR): It gives pilot distances from touchdown, heading instructions, altitudes relative to glide path and instructions on corrective action in the event that the aircraft is too high or too low. When the procedure terminates at two miles from touchdown, the distances from touchdown and altitudes relative to glide path checks are normally passed at one mile intervals. Where it terminates at less than two miles from touchdown, such checks are given each half mile. When pilot report runway in sight, it means that a landing will be effected; the radar approach may be terminated.

在精密雷达进近过程中,管制员向机组提供:

(1) 航向位置。

(2) 下滑道位置。

(3) 下滑道修正指令。

① 高于下滑道;

② 低于下滑道。

5. Check your minima:检查最低下降高度。

Minimum Descent Altitude(MDA): the lowest altitude, expressed in feet above mean sea level to which descent is authorized on final approach or during circle-to-land maneuvering in execution of a standard instrument approach procedure where no electronic glide slope is provided.

6. Possible reasons for heading change:

Positioning

Separation

Delaying

Identification

Obstruction

Traffic

Spacing

Sequencing

…

Exercises

I. Substitution practice.

A. CTL:CSN 3501, Shanghai Approach, squawk ident.

PIL:Squawking, CSN 3501.

— squawk 0123　　　Squawk ident

Unit Seven
Radar Control

— turn left heading 030 for identification   Left heading 030

B. CTL: CCA 982, turn right heading 340 to intercept the localizer, report established.

PIL: Right heading 340, CCA 982.

— turn left heading 270

— continue heading 330

— fly heading 310

— continue present heading

II. Listening practice.

Lesson 16-
Exercise 2

1. CSN 3194, for _____ turn right heading _____.
2. CSN 3690, turn _____ heading 150 for _____.
3. CSN 3172, _____ own navigation, position 40km _____ of VYK.
4. CES 2985, make _____ right for _____.
5. CSN 3430, _____.
6. CSC 8448, Wuhan Control, resume _____, direct to ZF, _____ track 220, distance 35 kilometers.
7. KAL 293, for _____, _____ 3 minutes en route, passing WUH at 12 _____.
8. CSN 3195, due to _____, make one _____ orbit in your present position and leave on heading 180.
9. PIL: Shanghai Control, OTRPP.

CTL: O-PP, report _____.

PIL: Level at _____ feet, heading 350, O-PP.

CTL: O-PP, _____.

PIL: Turning left heading 020, O-PP.

(Later)

PIL: O-PP, _____.

CTL: O-PP, _____, 15 miles south of _____.

10. PIL: Shanghai Control, D-LAMN, level at _____ m, heading 270.

CTL: D-MN, Shanghai Control, _____.

PIL: Turn right heading 300, D-MN.

(Later)

PIL: D-MN, _____.

CTL: D-MN not _____, resume own navigation.

PIL: _____, D-MN.

III. Translation.

汉译英

1. 国际1823,为了识别,左转航向两三洞。
2. 东方2587,雷达识别,VMB西南八公里,保持当前航向。

Lesson Sixteen
Radar Identification and Vectoring, Precision Radar Approach

3. 南方3357,雷达服务终止,恢复自主导航,直飞大王庄。
4. 国际1508,活动通报。不明活动,十二点钟方位,五公里,相对飞行,快速移动。
5. 南方3128,冲突解除,恢复自主领航,直飞大王庄。

英译汉

1. CCA 1326, identified, omit position report until VMB.
2. CSN 3308, make a three sixty turn left for delaying action.
3. CCA 1307, resume own navigation, direct CH, track 030, 20km.
4. CSN 3306, traffic information, unknown traffic 10 o'clock, 10 miles crossing left to right, fast moving.
5. CSN 3103, turn right heading 045, cleared ILS runway 13, report established.

Lesson Seventeen
Traffic Information and Radar Failure

Model Exchanges　　　　　　　　　　　　　　　　　　　　　　　Lesson 17

A.

CTL:BAW 301, unknown traffic 2 o'clock, 5 miles, crossing right to left, fast moving.

PIL:BAW 301, negative contact, request radar vectors.

B.

CTL:CCA 101, traffic information, traffic 11 o'clock, 7km, eastbound, B767, below you.

C.

CTL:CES 2289, traffic information, traffic 2 o'clock, 9km, westbound, B787, 900 meters above.

D.

CTL:JAL 019, turn right immediately heading 090 to avoid traffic 11 o'clock, 6km.

PIL:Right heading 090, JAL 019.

CTL:JAL 019, now clear of traffic, resume own navigation, direct to LKO.

PIL:Direct to LKO, JAL 019.

E.

CTL:All stations, traffic information, unmanned free balloon was estimated over MLJ at 1035, reported level 900m, moving west.

F.

CTL:CSN 3168, traffic 11 o'clock, 10km, converging, B767, 2400m, descending.

PIL:CSN 3168, looking.

CTL:CSN 3168, do you want vectors?

PIL:CSN 3168, negative vectors, traffic in sight, now passed clear.

Lesson Seventeen
Traffic Information and Radar Failure

G.

CTL: CSN 3871, radar contact lost due radar failure, converting to procedural control, reduce speed by 30kts to increase separation, report passing PEK.

PIL: Roger, CSN 3871.

Dialogue

A.

CTL: CSH 5320, Shanghai Control, traffic information, unknown traffic, 2 o'clock, 5 miles, crossing right to left.

PIL: Looking out, CSH 5320.

PIL: CSH 5320, negative contact, request vectors.

CTL: CSH 5320, avoiding action, turn right 30 degrees immediately and report heading.

PIL: Turing right 30 degrees, new heading is 090, CSH 5320.

CTL: CSH 5320, now clear of traffic, resume own navigation, contact Jinan Control on 123.75.

B.

CTL: All stations, Sierra Control, radar failure, radar control service terminated, establishing non-radar separation now.

CTL: CSN 3379, Sierra Control, report your position.

PIL1: Sierra Control, CSN 3379, TJK 49, 8900m, estimating XG 57.

CTL: CSN 3379, direct to XG, report passing XG.

PIL1: Roger, CSN 3379.

CTL: CSC 9281, make one orbit right for spacing. Break Break, DLH 290, climb and maintain 8900m, hold between LF and JR, right hand pattern, you're too close to the preceding aircraft.

PIL2: CSC 9281, roger, orbiting right.

PIL3: Roger, climb and maintain 8900m, hold between LF and JR, right hand pattern, DLH 290.

CTL: JAL 289, Sierra Control, traffic 11 o'clock, 7 miles, opposite direction, fast moving.

PIL4: Traffic in sight, now passed cleared, request descend to 7800m, JAL 289.

CTL: JAL 289, standby.

(Later)

CTL: All stations, Sierra Control, radar resume operation, I'll continue to pass instructions.

CTL: CSN 3379, squawk 3572 for identification.

PIL1: Squawking 3572, CSN 3379.

CTL: CSC 9281, holding ended, you may proceed en-route.

PIL2: Roger, CSC 9281.

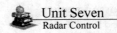

Unit Seven
Radar Control

CTL: JAL 289, cleared descend to 7800m, report reaching.
PIL4: Descending to 7800m, JAL 289.

New Words and Phrases

clear of 清除,离开,脱离
unknown *adj.* 未知的,不明的
unknown traffic 不明活动(冲突)
radar separation 雷达间隔
non-radar separation 非雷达间隔,程序间隔
aircraft proximity 航空器接近
radar clutter 雷达干扰
(compulsory) reporting point (强制)报告点
procedural control 程序管制
proceed en route 沿航路飞行

Notes

1. Traffic information includes:
时钟方位
飞行方向
相对距离
机型和高度(此项不清楚时用快慢表示)
2. 冲突航空器方位的表达方法:
11 o'clock
英语读法:eleven o'clock
汉语读法:十一点钟方位
3. Phrases about traffic information:
closing 接近
converging 汇聚
diverging 分散
crossing left to right 从左向右穿越
crossing right to left 从右向左穿越
overtaking 超越
opposite direction 逆向,相对飞行
parallel 平行
same direction 顺向
eastbound (westbound) 向东(西)飞行
unknown traffic 不明活动

132

Lesson Seventeen
Traffic Information and Radar Failure

slow moving 慢速移动
fast moving 快速移动

4. 当两架飞机距离太近危及飞行安全时,管制员会指挥飞机紧急避让(avoiding action);冲突解除后,管制员应尽早通知机组。

Exercises

I. Substitution practice.

A. CTL:CSN 3037,traffic 9 o'clock 6 miles crossing left to right,3000m.
— 12 o'clock,opposite direction
— 3 o'clock,parallel
— 11 o'clock,crossing left to right
— 2 o'clock,same direction

B. CTL:UIA 219,Beijing Control,radar failure,report your position.
— report present level
— report reaching 3000m
— report passing PEK
— report endurance

II. Fill in the blanks.

1. We have a radar failure,radar service _____.
2. You may _____ en-route.
3. All stations,Beijing control,radar now _____,resuming radar control.
4. Identification lost due radar _____.
5. CSN3021,traffic in sight,now _____.
6. UAL668,SSR resume operation,_____ 5312 for identification.
7. CSN301,_____ visual over the airport.
8. BAW 310,unknown traffic 12 _____,5miles,fast moving.
9. DLH 415,traffic 9 o'clock,9miles,_____ left to right,1800m,descending.
10. SAS 996,now clear of traffic,resume own _____,direct to OBY.

III. Listening practice.

1. CSN 3823,_____ contact,request _____.
2. EVA 087,_____,turn right 35 degrees immediately due to traffic.
3. JAL 091,turn left immediately heading _____ to avoid traffic 12 o'clock,_____ km.
4. SIA 528,Hongqiao Tower,unknown _____ sighted south of the field,slow moving,_____.
5. CTL:CSN 3503,traffic 3 o'clock,7 miles,_____,A320,9800m.
PIL:CSN 3503,_____,request vectors.

Lesson 17-
Exercise 3

133

Unit Seven
Radar Control

CTL: CSN 3503, descend to 8900m, report reaching.
PIL: _____ 9800m for 8900m, CSN 3503.
PIL: CSN 3503, reaching 8900m.
…
CTL: CSN 3503, _____, climb to 9800m.
PIL: Climbing, CSN 3503.

IV. Translation.

英译汉

1. JAL 021, Pudong Tower, unidentified flying object observed southwest of field, slow moving, proceed with caution.

2. CCA 1508, expedite descend to 8400m, I have a crossing traffic.

3. CSN 3379, radar failure, converting to non-radar separation, maneuver left and right to increase separation.

4. UIA 102, our radar is operation on low power at the moment, we are unable to spot primary targets, adjust speed so as to cross HG at 35.

5. CSC 9755, radar unserviceable, I can no longer assist you, reduce speed by 30kt to increase separation, report over SWK.

汉译英

1. 全体注意，二次雷达故障，一次雷达信号不稳定，自行保持间隔，小心飞行，雷达正在修理中。

2. 南京进近，南方3125，机载自动定向系统失灵，我们无法保持正确航迹，申请紧急援助。

3. 全日空988，你前方六十五公里的限制区处于激活状态，立即右转三十五度避让。

4. 南方3509，西安进近，活动通报，九点钟方位，七公里，正在超越，机型A320，高度三千六。

5. 雷达失效，报告燃油续航能力。

Unit Eight

Adverse Weather Information

Lesson Eighteen
Icing and Low Visibility

Lesson 18

Model Exchanges
A.
CTL: ACA 915, Nanjing Tower, have you got the trend?
PIL: ACA 915, negative, request the latest weather at Nanjing.
CTL: ACA 915, ceiling 300m, variable, visibility 1000m in fog, wind 350 degrees 5m/s.
PIL: No contact at minima, going around, ACA 915.
B.
PIL: Approach, UAL 105, the weather is deteriorating, a fog bank obscuring the north end of the field, the weather is below captain minimum, we would like to delay approach for 30 minutes.
CTL: UAL 105, roger, direct to the holding area and hold as published, expect approach at 1315.
C.
CTL: SAS 807, RVR RWY 35 touchdown 650m, mid-point 750m, stop end 700m.
PIL: SAS 807, roger, request radar vector for approach.
D.
CTL: CSN 3392, Hongqiao Tower, the weather is deteriorating, RVR runway 18R is less than 1500m, report intentions.
E.
PIL: Beijing Control, CSN 3392, we are encountering severe icing in cloud at 3600m, request climb.
CTL: CSN 3392, climb to 4800m, keep us advised if conditions continue or get worse.

137

Unit Eight
Adverse Weather Information

F.

PIL: Chengdu Control, SIA 158, our anti-icing system is inoperative, engine is low on power due engine starvation, request immediate descent.

CTL: SIA 158, descend and maintain 3000m, keep us advised if conditions continue or get worse.

G.

CTL: All stations, severe icing from FL180 upward.

Dialogue

A.

PIL: Bravo Tower, JAL 368.

CTL: JAL 368, go ahead.

PIL: JAL 368, outer marker.

CTL: JAL 368, you are number one, the weather is still deteriorating, RVR runway 18R is now less than 1500m.

PIL: Number one to land, JAL 368.

PIL: No contact at minimum, going around, confirm standard procedure, JAL 368.

CTL: JAL 368, climb straight ahead to 300m, then turn right to KD at 900m, contact Bravo Approach on 118.3.

PIL: Roger, climb to 300m and turn right to KD at 900m, contact 118.3, JAL 368.

B.

PIL1: Wuhan Control, AFR 821, 5400m maintaining, passing YIH 12, estimating HZ 28.

CTL: AFR 821, Wuhan Control, maintain 5400m on standard, report passing HZ.

PIL1: Roger, maintain 5400m, report passing HZ, AFR 821.

CTL: CSN 3106, hold over ZF, maintain 9200m due to flow control in Zhengzhou control area, expect further clearance in 20 minutes.

PIL2: Roger, hold over ZF, 9200m, CSN 3106.

PIL1: Wuhan Control, AFR 821, we have encountered severe icing condition, request lower level.

CTL: AFR 821, Wuhan Control, could you maintain present level for 3 minutes due to traffic?

PIL1: Negative, the leading edges are iced over, request immediate descent, AFR 821.

CTL: AFR 821, descend to 4800 meters, report reaching.

New Words and Phrases

variable *adj.* 变化的,不确定的,易变的

deteriorate *v.* 恶化,变坏

observe *v.* 观察；研究
encounter *v.* 遇到，遭遇
inoperative *adj.* 失效的
anti-icing system 防冰系统
starvation *n.* 饥饿
iced over 结满了冰
freezing level 结冰层（指结冰层下限）
keep sb advised 与某人保持联系，通知某人
obscure *v.* 使……模糊不清，掩盖
haze *n.* 霾
mist *n.* 轻雾
Fog is coming down 正在起雾
Fog is clearing up 雾在消散
precipitation *n.* 降水
drizzle *n.* 毛毛雨
tornado *n.* 龙卷风
squall *n.* 飑线
stratus *n.* （气象）层云

Notes

1. 低能见度处置程序：

(1) 能见度对飞机降落的影响比起飞的影响要大。当能见度较低时，应以跑道能见视程决定飞机能否降落。

(2) 机场以通播方式发布终端气象信息，管制员可以要求机场部门打开跑道灯以提高跑道视程。

(3) 如能见度过低，低于最低天气标准，或驾驶员在降落至决断高度时仍然无法看清跑道，管制员应指挥航空器复飞、改航或备降。

(4) 了解备降场和周边机场的天气情况。

2. anti-icing system inoperative：防冰系统失效。有关"失效"的表达方法还有 unserviceable，failure，out of work 等。

3. 多点观测的跑道视程分别代表接地点（touchdown）、中间段（mid-point）、停止端（stop end）的跑道视程。

4. Factors that might affect visibility：

(1) fog

(2) smog

(3) sand, haze, dust

(4) blowing snow

Unit Eight
Adverse Weather Information

(5) precipitation

(6) low stratus clouds

…

5. 结冰按照强度可分为严重（severe）、中度（moderate）和轻度（light）结冰三种。

Exercises

I. Substitution practice.

1. CTL: RVR runway 13 <u>1000 meters</u>.
— less than 800 meters
— greater than 1200 meters
— touchdown 600m, mid-point missing, stop end 700m

2. CTL: CSN 3302, the weather is below landing minima due to <u>dense fog</u>.
— a fog bank
— snowflakes
— a snow shower
— torrential rain
— dense drizzle

II. Translation.

汉译英

1. 南方3281，由于沙暴，天气正在恶化，跑道幺三跑道视程小于八百米，建议复飞。

2. 深圳塔台，东方5697，由于强降雨，天气低于机长最低标准，我们打算推迟进近，等待宝安机场天气好转。

3. 上航2896，武汉塔台，天气正在迅速恶化，正在起雾，并遮住了跑道洞六的一边，能见度不足七百米，请滑回停机位，我将与你保持联系。

4. 全体注意，武汉塔台，当前云底高三百米，预计二十五分钟内将升高到六百五十米，机场很快会重新开放。

5. 国际1472，云底高三百米，变化不定，能见度两千米，有雾，风两五洞，三米秒。

英译汉

1. Zhengzhou Control, KAL 866, engine is low on power due engine starvation, our leading edge is iced over, suspected air inlet also covered with ice, request immediate descent.

2. Guangzhou Control, CSN 3309, now clear of the icing area and closing back to the course.

3. Beijing Control, CES 5109, our yaw damper is inoperative, are we likely to run into severe icing on our route ahead?

4. SIA 195, Nanjing Tower, the weather is deteriorating, RVR runway 06 is less than 1500m, report intentions.

Lesson Eighteen
Icing and Low Visibility

5. CSN 3392, hold at Stand C5, we have already notified the relevant department, they will send a de-icing van for you immediately.

III. Listening practice.

1. JAL 305, RVR _____ touchdown _____ mid-point _____ stop end _____.

2. CPA 698, the _____ is deteriorating, _____ runway 31 is _____ 800m.

3. We have been _____ moderate _____, request _____ en route ahead.

4. We are encountering _____ in cloud at this _____, _____ system seems inoperative request _____ descent.

5. AFR 189, we have encountered _____ at this level, our anti-icing system is inoperative, request descent.

6. BAW 136, we have been notified by a Boeing 747 there is a _____ 40km ahead of you, _____.

7. UAL 164, expect _____ for you arrival.

8. Wuhan Control, CES 2940, _____ is inoperative and the leading edges are _____, request immediate _____.

9. PIL: Wuhan Control, JAL 390, we are _____ severe icing in cloud at 7,200m, request _____.
   CTL: JAL 390, _____, report reaching.

10. PIL: Wuhan Control, BAW 156, _____ system has _____ operation, request climb to 9,800m.
    CTL: BAW 156, Wuhan Control, cleared climb to 9,800m, _____.

141

Lesson Nineteen
Turbulence, Thunderstorm and Windshear

Model Exchanges

A.

PIL: Shenyang Control, CHH 7532, we've just been caught in severe turbulence, several passengers have been injured, would you check the type of traffic that's ahead of us?

CTL: CHH 7532, it was an A340, you may have flown into its wake turbulence, descend and maintain 8900m.

B.

CTL: CCA 101, reduce speed to 180kt, it should be sufficient to avoid running into the wake turbulence of the jet ahead of you (preceding aircraft).

C.

CTL: KAL 987, Guangzhou Control, thunderstorm is moving in your direction, go round the CBs by turning left 30 degrees, track out 25km from your present position.

PIL: Roger, KAL 987.

D.

CTL: CSN 3514, Beijing Control, climb to 11300m to overfly the CB, if unable, offset 25km left of track to circumnavigate it.

PIL: CSN 3514, can not climb 11300m due weight, going round the thunderstorm by left offset 25km.

E.

PIL: Hong Kong Approach, SIA 219, we have been struck by lightning, empennage damaged, left elevator lost, request priority landing and radar vector to final approach.

CTL: SIA 219, you are number one, radar will vector you until landing.

Lesson Nineteen
Turbulence, Thunderstorm and Windshear

F.

PIL: Beijing Control, CES 2289, request right offset to avoid the weather.

CTL: CES 2289, offset 3 miles right of track.

G.

CTL: CCA 101, advise if able to proceed parallel offset right 30km.

H.

CTL: CCA 918, Shanghai Approach, windshear warning, severe low level windshear observed in the vicinity of Pudong International Airport at 0700 hours, what is your intention?

PIL: CCA 918, we'd like to divert to our alternate airport.

I.

PIL: Hong Kong Tower, CSZ 8835, encountered strong windshear on final, gained 25kts between 600ft and 400ft followed by a loss of 30 knots between 400ft and the surface.

J.

CTL: All stations, caution, windshear reported by arriving B757 at 300m, 4km final, runway 36R.

Dialogue

A.

PIL1: Wuhan Approach, JAL 821, reaching 3600 meters on standard, we have an indication of weather 40km ahead of us on our present heading, request radar vectors to circumnavigate it.

CTL: JAL 821, Wuhan Approach, turn left heading 150, previous aircraft reported moderate turbulence and icing at your present level.

PIL1: Roger, left heading 150, request a lower level if possible, JAL 821.

CTL: JAL 821, standby.

PIL2: Wuhan Approach, CSC 8331, airborne at 23, climbing to 1500m on QNH, we encountered moderate turbulence during initial climb.

CTL: CSC 8331, Wuhan Approach, roger, radar contact, climb to 3000m on QNH, thunderstorm is approaching the local airport, it may cause turbulence.

PIL2: Climbing to 3000m on QNH, CSC 8331.

PIL1: MAYDAY, MAYDAY, MAYDAY, Wuhan Approach, JAL 821, electrical system failed after being struck by lightning, directional instruments unserviceable, maintaining 3600m on standard, heading 150, CBs passed clear, request landing priority.

CTL: JAL 821, roger MAYDAY. Break Break, all stations, stop transmitting, MAYDAY.

B.

PIL1: Wuhan Control, CCA 1508, we have just run into severe turbulence at 6600m, request descent to avoid it.

CTL: CCA 1508, negative due traffic, climb to 7200m, expedite until passing 6,900m,

Unit Eight
Adverse Weather Information

report reaching.

　　PIL1: Climb to 7200m, expedite until passing 6900m, call you reaching, CCA 1508.

　　PIL2: Wuhan control, JAL 367, we are caught in severe icing, windshield is iced over, anti-icing system is inoperative, request descent.

　　CTL: JAL 367, Wuhan Control, turn left 30 degrees, track out 20 kilometers before descending to 6000m. Caution, previous aircraft reported severe turbulence at 6600m.

　　PIL2: Turn left 30 degrees, 20km, descend to 6000m, JAL 367.

　　PIL1: Wuhan Control, CCA 1508, level at 7,200m, we have smoothed out now, thank you.

　　CTL: CCA 1508, maintain present level.

　　PIL2: Wuhan Control, JAL 367, reaching 6000m, the situation is getting better, we intend to divert to Wuhan.

　　CTL: JAL 367, roger, fly heading 160.

New Words and Phrases

　　wake turbulence (航空器)尾流
　　overfly v. 飞越
　　offset. v. 偏离(航线)
　　build-up 积云(通常指积雨云)
　　detour v. 绕道,绕飞
　　go round 绕开,绕飞
　　circumnavigate v. 绕飞
　　empennage n. (航空器)尾翼
　　rudder n. 方向舵
　　vicinity n. 附近
　　smooth out 变得平稳
　　directional instrument 方向仪表,助向仪表
　　yaw damper 偏航阻尼
　　up/down draught 上升/下降气流
　　CAT (Clear Air Turbulence) 晴空颠簸

Notes

　　1. 航空器遭遇雷暴时,管制员应及时通报机组雷雨发展趋势,了解相关机场的天气情况,了解机组绕飞意图,提供有关天气情况和绕飞建议。

　　2. 如果航空器遭遇长时间连续严重颠簸,应指挥机组上升或下降高度脱离颠簸区域。

　　3. (1)当机场附近有雷雨和风切变时,管制员应根据天气状况决定是否允许航空器起降,从安全角度出发,应指挥飞机改航或备降,并了解机组意图;

Lesson Nineteen
Turbulence, Thunderstorm and Windshear

（2）当航空器在航路上遭遇晴空风切变时，飞行员应立即向管制员报告风切变的强度、位置、范围和高度，管制员可以指挥该航空器改变高度层，如可能引起冲突，应调配周围航空器避让。同时，应向将要经过该颠簸区域的其他航空器发出风切变告警。

4. 颠簸按强度分为严重（severe）、中度（moderate）和轻度（light）颠簸三种。

5. 风切变按强度可分为严重（severe）、强（strong）、中度（moderate）和轻度（light）风切变四种。

6. 遭遇颠簸常用"be caught in, experience, encounter 或 run into turbulence"，遭遇雷击一般用"to be struck by lightning"。

Exercises

I. Substitution practice.

A. CTL：SIA 091, request flight conditions.
PIL：40km east of THE, 8400m, 1530 Zulu, continuous moderate turbulence between DFS and present position, SIA 091.

— encountering severe CAT
— moderate icing
— running into moderate turbulence
— experiencing severe turbulence 3 minutes ago, it has smoothed out now
— have just gone through severe icing

B. PIL：ANA 426, request clearance to go round the thunderstorm.

— deviate the thunderstorm from the right
— 20 degrees heading change right of track to detour the weather
— circumnavigate the CBs from the left
— 20 degrees offset to the right to go round the build-up.
— offset to the right from present position to avoid the CB
— climb to FL 411 to overfly the thunderstorm

Lesson 19-
Exercise 2

II. Listening practice.

1. Shanghai Control, QFA 102, we have an indication of weather _____ ahead of us, request clearance to _____ .

2. Hong Kong Approach, CSN 3175, we have been struck by lightning, _____, right elevator lost, request _____ to final approach.

3. Shanghai Control, CSN 3502, we have just been caught in _____, several passengers have been injured, we have to _____ .

4. Nanchang Control, CSH 2917, we have just been caught in severe turbulence, would you _____ .

5. Guangzhou Control, SIA 679, request _____ to avoid build-up.

6. DLH 831, Xiamen Control, thunderstorm moving in your direction, go round the CBs

145

Unit Eight
Adverse Weather Information

_____ from your present position.

7. SAS 136, Beijing Tower, hold at the holding point of runway 36R for a few minutes, the thunderstorm is _____.

8. CSH 2791, Shanghai Control, cleared to _____, report clear of CBs.

9. SAS 180, Beijing Control, _____, if not possible go round it 35 miles to the right.

10. SIA 915, cleared to divert to Dalian, you are number one, radar will assist you until landing, _____.

III. Translation.

英译汉

1. Nanjing Approach, CSN 3097, we encountered strong windshear at 200m, 5km final runway 24, nearly stalled.

2. Beijing Control, CES 5109, our yaw damper is inoperative, are we likely to run into turbulence on our route ahead?

3. Wuhan Control, CHH 8516, now clear of CB and coming back to the course.

4. Guangzhou Control, QFA 191, request alternate route, because last time we flew into turbulence, 2 engines feathered.

5. Guangzhou Control, JAL 405, engine's started again, seems to be running quite smoothly now, it flamed out when we were caught in updraught.

汉译英

1. 西安区域，日航295，当前航向前方四十公里显示有雷暴，申请从左侧绕飞。

2. 珠海进近，海南7153，起飞时遭遇了雷击，我们很难控制俯仰，升降舵可能受损，申请立即返航着陆。

3. 上海区域，深圳9175，高度拐两改平，遭遇严重晴空颠簸，有乘客受伤，申请改变高度避开颠簸区。

4. 兰州区域，大韩866，发动机马力不足，由于发动机进气受阻，我们的机翼前缘结满了冰，怀疑进气道也被冰覆盖，申请立即下降。

5. 白鹭8126，北京区域，前方四十五公里有雷暴，从你当前位置左转三十度出航二十五公里绕飞。

Unit Nine
Urgency and Distress

Lesson Twenty
Urgency

Model Exchanges

Lesson 20

A.

PIL: PANPAN PANPAN PANPAN, Hong Kong Approach, CAL 516, 30 miles west of Hong Kong at 9000 feet, a pregnant woman is about to give birth, request priority landing and medical assistance on arrival.

CTL: CAL 516, Hong Kong Approach, radar vectors for IGS approach runway 13, you'll be number one.

B.

PIL: PANPAN PANPAN PANPAN, Beijing Control, SIA 310, we've just been caught in severe turbulence, passengers have been hurt, request divert to Zhengzhou for landing, WXI 1105 at 7200m.

CTL: SIA 310, Beijing Control, roger, cleared divert to Zhengzhou, contact 122.20 for Zhengzhou Control.

C.

PIL: Xi'an Ground, CSH 2156, a male passenger refused to power off his mobile phone and had a fight with our flight attendant, situation is now under control, one flight crew got hurt in the underbelly.

CTL: CSH 2156, roger, report your intention.

PIL: Request taxi back to the apron and security assistance, CSH 2156.

D.

CTL: ANA378, I'm unable to give you further clearance at this time, I've got a problem, the radio has failed at ACC, if you read, squawk 7600, I say again, 7600, call you back.

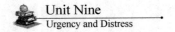

Unit Nine
Urgency and Distress

CTL: ANA378, squawk observed, will continue to pass instructions.

E.

CTL: CSN 3392, if radio contact lost, maintain 900 meters on QNH, direct to AA217, and hold as published.

F.

PIL: Zhengzhou Control, DLH 718, transmitting blind due to receiver failure, DLH 718, over ZHO at 9500m, ETO WXI 45; transmitting blind, DLH 718, over ZHO at 9500m, ETO WXI 45, next transmission at 55.

G.

PIL: PANPAN PANPAN PANPAN, Beijing Approach, CCA 1589, intercepted urgency call from JAL 813, unable to contact you due receiver failure, 185 passengers plus six crew members on board, request emergency landing at Beijing, position 35km south at 2700m.

CTL: CCA 1589, roger.

Dialogue

A.

PIL1: Shanghai Approach, ANA 306, we have just found a note on board saying a bomb has been planted in the cargo hold, request landing priority and emergency assistance, please inform ground we intend to execute emergency evacuation after landing.

CTL: ANA 306, Shanghai Approach, roger, descend to 900m, fly heading 310, vectoring for ILS approach runway 36R, you are number one.

PIL1: Descend to 900m, heading 310, runway36R, number one, ANA 306.

CTL: ANA 306, turn left heading 300 for downwind.

PIL2: PANPAN PANPAN PANPAN, Shanghai Approach, CDG 6491, airborne weather radar malfunctioned, request return to land, we need mechanic assistance upon arrival.

CTL: CDG 6491, Shanghai Approach, roger PANPAN, turn right, heading 240, descend to 1,200m, vectoring for ILS approach runway 36R.

B.

PIL1: Beijing Tower, CES 2537, airborne 25, 200m climbing.

CTL: CES 2537, climb on runway heading to 1200m on QNH, contact Approach on 119.7.

PIL1: Runway heading to 1200 m on QNH, 119.7, CES 2537, good day.

PIL2: PANPAN PANPAN PANPAN, Beijing Tower, CHH 7581, we are returning, we are having trouble retracting the nose gear due hydraulic system failure, request priority landing at Beijing Capital International Airport.

CTL: CHH 7581, roger, descend to circuit altitude 600m, join downwind of runway 18L, you are number one, report final.

PIL2: Descending to 600m, join downwind of runway 18L, CHH 7581.

150

PIL3: Beijing Tower, JAL 309, approaching holding point runway 18L, request line up.

CTL: JAL 309, hold short of runway 18L. Break Break, CDG 9532, orbit right from your present position, standby for further clearance, PANPAN in progress.

PIL3: Roger, holding short of runway, JAL 309.

PIL4: Roger, orbiting right from present position, CDG 9532.

New Words and Phrases

 jettison *vt.* 丢弃,放油
 divert *v.* 改航
 squawk *n.* 应答机
 engine flame out 发动机熄火
 engine feathered 发动机顺桨
 hijack *v.* 劫持
 bird ingestion 鸟吸
 plant *v.* 安装
 airborne *adj.* 机载的
 malfunction *v.* 失灵
 hydraulic *adj.* 液压的

Notes

1. PANPAN PANPAN PANPAN—indicates an urgent condition, one of being concerned about safety, and requiring timely BUT NOT immediate assistance.

 紧急信号表示一种紧急情况,涉及航空器、其他车辆和机上人员安全,需要采取及时但不是立即援助。

 2. 紧急电文的格式:

 (1) PANPAN(三遍)

 (2) 管制单位呼号

 (3) 本机呼号

 (4) 应急情况的性质

 (5) 机长意图

 (6) 航空器位置、高度及航向

 (7) 其他情报

3. 当收到机组的紧急信号后,管制员可以用以下句型询问或证实:

 (1) report your intention 报告意图

 (2) understand your situation 明白你的处境

 (3) What is the nature of your emergency? 你是什么性质的紧急情况?

Unit Nine
Urgency and Distress

4. 当航空器由于接受机故障不能与管制单位建立联系时，应按照规定时间或位置在原来的频率上发送电文，电文前加上"Transmitting blind due to receiver failure"。电文内容应重复一遍，并通报下次发报时间。

Exercises

I. Substitution practice.

A. PIL: PANPAN PANPAN PANPAN, Beijing Approach, KAL 939, <u>we are returning, smoke detected in the rear cargo hold</u>, position 20km north of VYK heading 340 at 5100m.

— No. 1 engine feathered
— engine runs rough due bird ingestion
— airborne weather radar shut off, reason unknown
— security problem

B. PIL: PANPAN PANPAN PANPAN, Beijing Control, CSN 3265, <u>an old woman passenger has fainted</u>, we have to divert to Tianjin, 50km east of VYK at 6000m.

— several passengers got injured after severe turbulence
— a pregnant woman is about to give birth
— a young lady has a breathing problem, she is in immediate danger
— running into severe icing

II. Listening practice.

1. PIL: Tower, CSN 3586, take-off _____ due burst tyre.
   CTL: CSN 3586, vacate runway next left.
   PIL: CSN 3586, we are standing on the stopway, _____ emergency _____, request emergency service.
   CTL: CSN 3586, _____ will be available in a few minutes.

2. PIL: Wuhan Approach, CCA 1038, we have to _____ because of landing weight requirements dictated by structural _____ of the aircraft.
   CTL: KAL 012, set heading to MN, after passing MN heading 270 to start _____, upon _____ turn right direct to BX, maintain 5100m.

3. CSN 3102, transmitting blind due _____ failure, CSN 3102, VYK 6000m descending for ILS runway 36R, CSN 3102, 6000m descending for ILS runway 36R, next _____ DALUE.

4. Pudong Tower, UPS 633, fire is _____ now, but we request _____ on landing.

5. Dalian Control, MDA 271, port engine air intake _____ birds during climbing and engine _____.

6. CTL: CSN 3981, Beijing Control, confirm squawk.
   PIL: Beijing Control, four _____ attempted to _____ our aircraft.
   CTL: CSN 3981, what's their intention?

152

PIL: The hijackers _____ us to fly to Hong Kong, and threatened to blow up the aircraft if we fail to comply with their requirement.

7. CTL: CHH 6303 _____ lost, if you read, squawk 7600, I say again 7600.

CTL: CHH 6303, squawk _____, will continue to pass instructions.

8. Shanghai Control, AIC 204, it looks like our _____ is iced over. Does the MET _____ severe icing in the area?

9. Xiamen Tower, DLH 278, first officer windshield was _____ due unknown foreign object _____ .

10. PIL: Dalian Tower, ANA 399, we have to make a belly landing, request _____ the runway and _____ instructions.

11. PIL: Beijing Tower, SIA 406, we'll have to make a _____ landing, request _____ at the touchdown zone of the runway.

PIL: Please have the _____ equipment stand by.

CTL: SIA 406, Beijing Tower, runway is being _____, expect approach shortly.

Ⅲ. Translation.

英译汉

1、Guangzhou Control, UAL 198, we're having difficulties in maintaining straight course, the aircraft tends to drift to the right, rudder probably malfunctioned.

2. Guangzhou Approach, CCA 1898, overheat lights flashed on, the oil needs cooling down.

3. PANPAN PANPAN PANPAN, Shanghai Control, CSN 3509, No. 3 engine severe vibration, reason unknown, we may shut down the engine anytime.

4. PANPAN PANPAN PANPAN, Chengdu Control, JAL 105, dense smoke coming out of No. 4 engine cowling, fire suspected.

5. Chengdu Approach, CPA 508, outer windshield panel appears to be crazed by hailstones or lightning, because we heard a loud thump.

汉译英

1. PANPAN PANPAN PANPAN, 哈尔滨区域, 上航5182, 高度表出现故障, 申请引导我们下降高度到云下。

2. 深圳塔台, 山东8291, 由于液压系统问题, 刹车和前轮转向系统失效。

3. 大连区域, 北欧583, 液压系统仍漏油, 但能够操纵飞机到北京。

4. PANPAN PANPAN PANPAN, 广州进近, 深圳8829, 截听到东方2837的紧急讯号, 低油量, 申请优先着陆, 位置在机场西北十公里, 高度三千。

5. 广州进近, 国际1307, 到达深圳时需要一位医生, 因一名机组成员感觉不适, 收到电报后, 请转告国航。

Lesson Twenty-one
Distress

Model Exchanges
A.

PIL: MAYDAY MAYDAY MAYDAY, Nanjing Approach, SIA 074, we have been struck by lightning when climbing to 4200m, the rear fuselage is damaged, we've lost control of the rudder and left elevator.

CTL: SIA 074, Nangjing Approach, roger MAYDAY.

B.

PIL: MAYDAY MAYDAY MAYDAY, Shanghai Approach, DAL 340, our low fuel pressure warning light is on and we have only 30 minutes fuel left, request priority landing at Pudong International Airport.

CTL: DAL 340, Shanghai Approach, roger MAYDAY, turn right heading 200 and descend to 1500m, runway 17, you are number one in traffic.

C.

PIL: MAYDAY MAYDAY MAYDAY, Tango Control, CPA 875, we have experienced a hydraulic failure and are in difficulty.

CTL: CPA 875, Tango Control, squawk 7700, howcan I help you?

PIL: Request radar vectors toward the aerodrome while we assess the situation, CPA 875.

CTL: CPA 875, turn left heading 250 for a vector to CHG, maintain 9800 meters, advise when ready for further clearance.

D.

PIL: MAYDAY MAYDAY MAYDAY, Zhuhai Approach, BAW 942, cockpit on fire, making forced landing 25km west of Shenzhen, passing 2400m heading 130.

CTL:BAW 942,Zhuhai Approach,roger MAYDAY.

E.

PIL:MAYDAY MAYDAY MAYDAY, Beijing Control, CCA 982, cabin altitude rising rapidly,emergency descent to 3300m,heading 230.

C:CCA 982, Beijing Control, roger MAYDAY. Break Break, all stations, emergency descent between PSN and VMB,all aircraft below 9800m between PSN and VMB leave A593 to the east immediately.

F.

PIL:Beijing Tower,DLH 169,we're having a full emergency now,we're unable to extend the nose gear,we'd like to make a low pass so that you can check whether nose wheel is down.

CTL:DLH 169,cleared low pass,runway 18R,not below 100m,report final.

G.

CTL:All stations,Wuhan Tower,stop transmitting,MAYDAY.

PIL:Wuhan Tower,BAW 301,cancel distress,engine restarted,runway in sight,request landing clearance.

CTL:BAW 301,runway 22,cleared to land.

PIL:Cleared to land,BAW 301.

CTL:All stations,Wuhan Tower,distress traffic ended,out.

Dialogue

A.

PIL1:Wuhan Control,CXA 8122,over PLT,7500m,estimating KHN 0129.

CTL:CXA 8122,roger,report over KHN.

PIL2:MAYDAY MAYDAY MAYDAY,Wuhan Control,CES 2361,the instrument indicates that our cabin altitude is too high,suspected cabin depressurization,request immediate descent.

CTL:CES 2361,Wuhan Control,roger MAYDAY,turn right 30 degrees,track out 20km, then descend and maintain 3900m.

PIL2:Wuhan Control, CES 2361, we are executing emergency descent due rapid decompression.

CTL:CES 2361, roger. Break, Break, CXA 8122, turn right heading 270 immediately to avoid the emergency descending traffic.

PIL1:Roger,right heading 270 immediately,CXA 8122.

CTL:CES 2361,report your intention.

PIL2:Wuhan Control,our cabin altitude is 2000m and our destination is Wuhan,but we intend to divert to Nanchang to have a check,CES 2361.

CTL:CES 2361,roger,cleared divert to Nanchang,maintenance staff will be ready for you upon landing.

Unit Nine
Urgency and Distress

B.

PIL: MAYDAY MAYDAY MAYDAY, Nanjing Approach, CSZ 9127, we have lost power on the left engine.

CTL: CSZ 9127, Nanjing Approach, roger MAYDAY, what's your intention?

PIL: CSZ 9127, request radar vector to Nanjing Lukou International Airport, runway 06 for landing and emergency assistance.

CTL: CSZ 9127, roger, turn right heading 030 for a vector to ILS approach of runway 06, descend and maintain 4800 meters.

PIL: Right heading 030, runway 06, descend and maintain 4800 meters, CSZ 9127.

PIL: Nanjing Approach, CSZ 9127, 4800 meters, left engine seemed to resume operation, request priority landing in case something else should happen, please have the emergency assistance and mechanics standby.

CTL: CSZ 9127, roger, cleared straight-in ILS approach runway 06, you are number one, emergency assistance is standing by, we will inform the airport maintenance for you.

New Words and Phrases

  fuselage *n.* 机身
  rudder *n.* 方向舵
  elevator *n.* 升降舵
  hydraulic *adj.* 液压的,水力的
  execute *v.* 执行,实施
  evacuation *n.* 撤离,疏散
  aisle *n.* 过道,通道
  decompression *n.* 失压
  depressurization *n.* 失压
  corridor *n.* 走廊,通道
  supercharger *n.* 增压器
  shatter *v.* 砸碎,粉碎
  ditching in sea 海上迫降,水上迫降

Notes

1. MAYDAY MAYDAY MAYDAY—indicates that an aircraft is being threatened by serious or imminent danger and requires immediate assistance.
遇险信号表示航空器正遭受严重的或迫近的危险威胁,需要立即援助。
2. 遇险电文的格式:
(1) MAYDAY(三遍)

Lesson Twenty-one
Distress

（2）管制单位呼号

（3）本机呼号

（4）遇险情况的性质

（5）机长意图

（6）航空器位置、高度及航向

（7）其他情报

3. 应答机编码：7500——劫机；7600——通讯失效；7700——遇险

4. 强制无线电静默：在航空器遇险呼叫后之后，空中交通管制会让其他在该频上通话的航空器保持缄默——强制无线电静默，这种强制静默有时可以针对某一架航空器，如"CES 981, stop transmitting, MAYDAY"。

5. 解除险情：航空器在解除险情后应及时发送取消该情况的电文，如"All stations, Tango Control, distress traffic ended."。

6. 静默终止：当空中交通管制单位收到遇险航空器解除遇险的报告后，向在该频率上的所有航空器发出静默终止指令。

7. 航空器紧急下降应注意两种情况：

（1）如在紧急下降之前时间允许的话可通知管制单位以获得空中交通管制同意；

（2）在来不及通知管制单位的紧急情况下，驾驶员可立即采取下降措施并报告管制单位，空中交通管制在收到紧急下降报告后，会立即采取相应的措施来保证其他航空器的安全，让其他航空器避开紧急下降航径或区域。

Exercises

I. Substitution practice.

Example

A. PIL: MAYDAY MAYDAY MAYDAY, DLH 699, take-off aborted, <u>engine on fire</u>, request emergency assistance.

— generator failure

— fuel boost pump in right main tank is inoperative

— de-icing system is unserviceable

— rear cabin door opened in flight

— number one engine flamed out

B. PIL: MAYDAY MAYDAY MAYDAY, Beijing Control, CSN 981, we lost control of the aircraft, <u>pitch control inoperative</u>, 9800m, descending.

— decompression

— flaps are jammed

— we've lost the rudder and right elevator

— left aileron failure

— the aircraft is rolling right and left up to 20 degrees

157

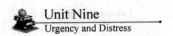

Unit Nine
Urgency and Distress

II. Fill in the blanks.

1. PIL: MAYDAY MAYDAY MAYDAY, JAL 310, smoke coming from engine bleed air system, we suspect that something is on _____, we have dropped the _____ mask.

CTL: JAL 310, you're cleared to descend at your own _____.

PIL: We're examining the mechanism and trying to locate the fire, request _____ landing.

CTL: JAL 310, cleared _____ approach, RWY 18R, you'll be number one.

PIL: Straight-in approach, RWY 18R, JAL 310.

2. PIL: MAYDAY MAYDAY MAYDAY, ANA 121, we've been _____ by an unknown foreign object, caused a big hole on the fuselage, one passenger has been _____ out.

CTL: ANA 121, roger MAYDAY, reduce speed and descend to 3600m, report reaching. Break break, all stations, stop _____, MAYDAY. Break, all stations, _____ descent between TER and VNG, all aircraft below 8400m between TER and VNG leave the area to the east immediately.

PIL: Beijing Control, ANA 121, _____ at 3600m.

III. Listening practice.

1. PIL: QFA 384, Kunming Tower, we have cut off No. 1 engine due to a fuel line _____.

CTL: QFA 384, join _____ RWY 36R, _____ on your own discretion.

2. PIL: MAYDAY MAYDAY MAYDAY, Hong Kong Approach, CSN 3017, port engine failed, will attempt to land at your airport, _____ west of Charlie, 8900ft, heading _____.

CTL: CSN 3017, Hong Kong Approach, roger MAYDAY, radar _____ RWY 13, turn right direct to CH and descend to 6900ft, you'll be number one.

3. PIL: Shanghai Ground, CES 2913, a passenger refused to _____ his cell phone and had a fight with our flight attendant, situation is now under _____, one flight crew got hurt, request taxi back to the apron and _____.

4. CTL: Aircraft _____ 7600, Beijing Approach, squawk standby.

CTL: If landing Beijing Airport, squawk _____.

CTL: Ident observed, expect ILS runway 36L approach, wind 310 degrees at 4 meters per second, QNH 1002. If _____, squawk ident.

5. PIL: Tianjin Tower, MDA 471, for the _____ system problem, we have lost the _____, we'll try _____ extension.

IV. Translation.

英译汉

1. Shanghai Control, CCA 1279, we've been hijacked, the guy wants us to divert to Manila for landing, but we'll have to refuel at Pudong airport, request immediate descent.

2. MAYDAY MAYDAY MAYDAY, Beijing Approach, CSN 3158, a bomb exploded on

board and the cabin is on fire, the aircraft is disintegrating, making forced landing 20 miles south of VYK.

3. Dalian Control, SIA 311, port engine air intake ingested birds during climbing and engine flamed out.

4. All stations, emergency descent between VY and WX, all aircraft below 3000m between VY and WX leave corridor to the east immediately.

5. Chengdu Approach, CES 2813, we are running short of fuel because we've been holding for one hour due flow control, request immediate landing at Chengdu Shuangliu airport.

汉译英

1. MAYDAY MAYDAY MAYDAY,北京进近,南方3309,二发熄火,由于不明外物撞击,怀疑吸入飞鸟,申请优先着陆及紧急救援服务。

2. 济南区域,全日空839,我们正在上升,左发整流罩部分脱落,击中平尾,我们很难控制飞机俯仰,申请返航及优先着陆。

3. 北京区域,汉莎734,取消险情,航空器恢复控制,我们打算备降郑州。

4. MAYDAY,MAYDAY,MAYDAY,武汉进近,立荣178,我们在上升过程中被一只大秃鹫撞击,副驾驶风挡破裂,我们有释压的危险,请指示。

5. 沈阳进近,南方3085,我们计划返回落地,但是由于航空器结构限制对着陆重量有要求,我们需要首先放油二十吨。

Key to exercises

Lesson One

II. Give the abbreviated forms of the following call signs.
DMH   BPD   AFR 901   CDG 573   Shorts AC   Cessna IG
BAW PC   ABB CD   N582   Q741

III. Dictation: Write down what you hear on the recording.
1. 35   158   694   700   2300   14832   1800   26954   196319   10200   3600
2. 2700m   4200m   6900m   11600m   14900m   8900m   10700m
3900ft   9800ft   FL187   FL256   FL430   FL351   FL207
3. 118.3   119.23   120.34   123.76   119.05   122.37   124.39
4. 1423   0130   0945   2216   1918   1307   2356   0438
5. ZX   HY   FRT   AOP   QSK   IUYT   CVBN   QASW   GFHIX   MNBVC   QRYIL   AERFHU   OLPUMNE   WXCGFTYM   KJHGFDSANI

Lesson Two

II. Fill in the blanks.
1. go ahead
2. station calling/say again
3. contact/117.2
4. correction
5. monitor
6. how do you read

Key to exercises

7. confirm
8. disregard
9. standby
10. approved

III. Dictation: Write down what you hear on the recording

1. All stations, G-ABCD, leaving 9500 meters for 7500 meters.
2. B-KGIL, contact Charlie Control on 117.45.
3. N376217, standby 123.25 for Tower.
4. CCA 1682, monitor Bravo Tower on 119.3.
5. Ground, G-EFCD, request departure information.
6. CSN 3527, Hongqiao Ground, departure runway 36R, wind 330 degrees, 8m/s, QNH 1012, temperature 23, dew point 20, RVR 750 meters.

Lesson Three

I. Listen to the following ATIS or VOLMETS, write down as much information as you can.

1. This is Paris-Charles de Gaulle information M recorded at 1800 Z, landing runway 27, departure runways 26R/26L. Transition level 60,. wind 300 degrees, 9 knots, visibility 10km or more, ceiling scattered at 12000ft, temperature 24, dew point 11, QNH 1018. Advise controller you have information Mike at initial contact.

2. Beijing Capital Airport, ATIS Information P, 2103Z, landing runway 18L. Departing runway 18R, expect ILS approach, wind 120 at 6 meters per second, visibility 2400 meters, cloud, broken 500 meters, temperature 20, dew point 18, QNH 1011. On initial contact, advice you have information P.

3. Hongqiao International Airport information K. 0700 zulu. Main landing runway 36R, ILS approach, main departure runway 36L, runway covered with thin patches of rime, braking action poor, wind 280 degrees at 6 meters per second, gusting 12 meters per second, visibility 4 kilometers, light freezing rain, overcast, ceiling 900 meters, temperature minus 2, dew point minus 5, QNH 1006 hectopascals. Taxiway L closed. Advise on initial contact you have information K.

4. Xi'an Xianyang International Airport Met report 1645 Zulu. 170 12m/s, visibility 2000m, fog patches, 3 octas 600m, temperature plus 5, dew point missing, QNH 1021, no sig

5. Beijing Capital International Airport VOLMET 0900 GMT. 210 19kts, visibility 3500m, haze, 7 octas 150m, temperature −2, dew point −4, tempo, visibility 2000m, freezing rain, out.

II. Dictation: Write down what you hear on the recording.

1. Snow and ice on runway 23

161

Key to exercises

2. Runway 18 covered with patches of dry snow

3. Loose snow on runway 06

4. Wet snow on runway 23

5. Compacted snow on runway 07

6. Firm snow on runway 18

7. Slush on runway 27

8. Frozen slush on runway 12

9. Ice patches on runway 10R

10. Snowdrifts on right side of runway 17

11. Frozen ruts and ridges on grass strip

12. Snow removed length:900m,width:30m

Ⅲ. Translation.

汉译英

1. CDG 8592,for your information,there is a tug broken down near taxiway B,we will send you a Follow-me car.

2. CCA 1809,threshold of runway 17 is contaminated by air fuel,threshold displaced by 175 meters,land with caution.

3. KAL 002,ATIS at the local airport is unserviceable,request aerodrome information from Ground when ready for departure.

4. Tower,CYZ 9123,taxiway B is very slippery,our left main gear is stuck in the mud, request assistance.

5. SIA 109,cancel your flight plan,airport is temporarily closed,storm is coming.

英译汉

1. 南方3518,哈尔滨塔台,我们三小时前有个紧急情况,跑道铺设了泡沫毯,机场将在二十分钟后开放。

2. 国泰576,武汉进近,预计进场时能见度低,与我们保持联系。

3. 上航2734,跑道洞四部分覆盖薄冰,十分钟前EMB 幺四五报告落地时刹车效应差。

4. 附近已经下了一阵子雪,但跑道上没有积雪的报告。

5. 云底高不久将上升到一千米,预计机场不久将重新开放。

Lesson Four

Ⅱ. Translation.

汉译英

1. Beijing Ground,CCA 1587,radio check 129.0,how do you read?

2. CSN 3064,I read you 2,loud background noise,adjust your transmitter.

3. Shanghai Ground,CES 5106,destination Beijing with information B,ready to copy ATC

Key to exercises

clearance.

4. CSN 3392, cleared to Guangzhou via flight planned route, runway-in-use 07, follow 01 departure, initial altitude 900m on QNH 1012, cruising level 9800m, request level change en route, squawk 5124, when airborne, contact Approach on 121.1.

5. Station calling Pudong Tower, say again your call sign.

英译汉

1. 日航 104,香港地面,你的信号不稳定,检查你的发射机,给我一个短数。

2. 南方 3398,可以沿飞行计划航路放行到香港,巡航高度幺洞拐,沿 Delta 洞八标准程序离场,应答机 3542。

3. 大韩 761,北京地面,风两两洞,八米秒,温度十二摄氏度,露点九摄氏度,QNH 1021,离场跑道三六左。

4. 东方 5128,你的复诵不正确,确认应答机 6123。

5. 由于接收机故障,我听不到你。

III. Fill in the blanks.

1. destination/departure

2. Delivery/DTV

3. cleared

4. Station/call sign

5. adjust (check)

6. radio check

IV. Listening practice.

1. readability/transmitter

2. radio check

3. destination/ATC clearance

4. gusting/dew point/departure runway

5. say again/broken

Lesson Five

II. Translation.

汉译英

1. Hong Kong Ground, CSN 3704, stand 07, information B, request push-back and start-up.

2. SIA 617, slot time 45, start up at own discretion.

3. Ground crew, please confirm number 2 engine is rotating.

4. Engine start completed, please disconnect interphone.

5. CXA 8194, your slot time will be delayed for 40 minutes due to flow control, push back and start up at own discretion.

163

Key to exercises

英译汉

1. 南方3503,北京地面,预计延误十分钟,由于冲突。
2. 日航701,一架波音747正从你后面穿过,稍后给你推出指令。
3. 美联航134,停机位Alpha九,通播Charlie,申请开车去深圳。
4. 四川8401,预计离场五洞分,开车联系你。
5. 发动机全部启动结束,检查一切正常,请撤轮挡和其他地面设备。

Ⅲ. Fill in the blanks.

1. push-back
2. approved
3. released/set
4. discretion
5. departure
6. confirm
7. rotating/running
8. delay
9. disconnect
10. clear

Ⅳ. Listening practice.

1. start-up
2. delay/baggage identification process
3. boarding/completed
4. catering/slot time
5. push-back/replacement
6. facing north
7. started/ground equipment
8. expect/handling
9. my apologies/failed/approved
10. engine start
11. Guangzhou Ground, JAL 137, Gate 3, request push-back.
12. CCA 1537, expect 10 minutes delay due to vehicle broken down, monitor 118.7.
13. CSN 3381, good morning, slot time 30, start-up 5 minutes before.
14. Engine start completed, please disconnect interphone.
15. KAL 410, expect departure 55, start-up at your own discretion.

Lesson Six

Ⅱ. Translation.

汉译英

1. DLH 076, taxi via taxiway M and C to holding point runway 36L, taxi with caution.

Key to exercises

2. JAL 782, cleared to cross runway 24, report vacated.

3. Ground, I'm not familiar with the local airport, please send me a Follow-Me car to guide me to the runway-in-use.

4. DAL 655, hold short of intersection V, give way to B757 passing from right to left.

5. Ground, SAS 124, stand D2, request detailed taxi instructions for runway 32.

英译汉

1. 英航 024，下一个道口右转脱离跑道，跟着你前面的航空器滑行，我将与你保持联系。

2. 日航 521，立即停止滑行，前方道口被一辆故障车阻塞。

3. 新航 094，沿滑行道 Bravo 三和 Alpha 四滑到北停机坪，停机位 Romeo 三，目视引导报告。

4. 国际 1562，注意，波音 747 正从右侧超越，他是第一个离场。

5. 英航 386，在跑道洞三上调头，沿滑行道 Alpha 幺脱离跑道。

III. Fill in the blanks.

1. hold at/approaching
2. hold/way
3. backtrack
4. short of
5. via
6. straight
7. line up and wait
8. advised
9. cross
10. expedite

IV. Listening practice.

1. vacated/terminal/taxiway centerline lighting
2. landing MD-11/wait
3. pull in/traffic in sight
4. heavy/ready for taxi
5. straight ahead/second turning on the left
6. left to right
7. south terminal/threshold
8. follow/yellow/hangar
9. one eighty/backtrack/fire engine
10. work in progress

V. Look at the following pictures and give proper instructions

a. KTD, take the second right.

165

Key to exercises

  b. KTD, hold short of the intersection, give way to the aircraft taxiing in.

  c. KTD, taxi with caution, GHL overtaking you on your right.

  d. KTD, pull in to the left to holding point K, give way to GHL passing on your right.

  e. KTD, expedite taxi, GHL is taxiing behind.

  f. KTD, hold short of the intersection, GHL is taxiing out.

Lesson Seven

Ⅰ. Translation.

英译汉

1. 深圳8827，直线上升到海压高度九百，然后右转，可以起飞。
2. 华航905，跑道外等待，看到五边的麦道幺幺报告。
3. 山东9543，取消，我重复一遍，取消起飞，跑道上有施工人员。
4. FDX 528，广州塔台，立即起飞，否则跑到外等待，长五边有波音777。
5. 南京塔台，汉莎198，起飞中断，由于爆胎，右主轮陷在泥里，申请客梯车和摆渡车将旅客送到候机楼。

汉译英

1. CHH 6428, Xi'an Tower, unable to issue take-off clearance at this time due military activity, standby for further clearance.

2. SIA 512, line up, be ready for immediate departure.

3. KLM 793, behind A380 on short final, line up and wait.

4. ANA 101, hold position, cancel take-off, I say again, cancel take-off, suspected SARS case reported on board.

5. CCA 1503, after departure, climb on runway heading to 900 meters on QNH, then turn left, follow SID, cleared for take-off.

Ⅱ. Fill in the blanks.

1. ready/departure
2. vacated/up/wait
3. take-off/take off
4. cancel
5. cleared
6. final/sight

Ⅲ. Listening practice.

1. holding point/wait/problem/nose wheel steering/jammed/tug/tow
2. for a few minutes/thunderstorm/far end
3. reaching/parking stand/brakes/overheating/holding bay/convenient left turn

Key to exercises

Lesson Eight

I. Fill in the blanks.

1. to or destination/departure/departure RWY/gusting /QNH/QNH 1008/CSN 3051
2. JAL 306/request departure information/departure runway 16/QNH 1012/JAL 306
3. reaching/continue/set
4. request/climb/contact

II. Listening practice.

1. JAL 905, change to Foxtrot Control on 126.9

go ahead

JAL 905, turn right heading 130, climb to FL217, expedite until passing FL157

JAL 905, report reaching FL217

Roger, JAL 905, contact Delta Control on 128.2

2. AAL 108, change now to November Control on 127.3

AAL 108, go ahead

turn right heading 230 and climb to FL236, report reaching

Roger, AAL 108, climb to FL276.

Change now to Alpha Control on 129.5.

IV. Translation.

英译汉

1. 国际1035,北京进近,在大王庄之前上升到五幺。
2. 东方5287,香港进近,上升率不小于十五米秒。
3. 西安进近,南方3579,我们刚才遭遇严重颠簸,我们前方是什么型号飞机？
4. 达美057,由于冲突,你能否调整上升率,在到达FIR边界之前上升到高度层190之上？
5. 成都进近,日航762,我们正在返航,轮舱好像失火,告警灯刚才亮了,申请优先着陆和紧急服务。

汉译英

1. EVA 813, turn left heading 245 to intercept the 270 radial of HL VOR.
2. UIA 639, climb to reach 8400 meters at LIG, expedite until passing 7200m.
3. CPA 799, cancel SID, turn right direct to KDL, climb and maintain 6300m.
4. BAW 134, follow B09 departure, climb to 7200 meters, report passing 6300 meters.
5. CXA 9816, turn left immediately toavoid colliding with an unknown conflicting traffic.

Lesson Nine

II. Listening practice.

1. depressurization/emergency descent/emergency landing

Key to exercises

2. estimating KG at 28/KG/the weather ahead of you/severe vibration on the right side

3. caught in severe turbulence

medical assistance on landing

20 kilometers from PIN

ambulances

suffering from minor bruises

Ⅲ. Translation.

英译汉

1. 武汉区域，东方5183，KVN四五分，高度幺洞四，预计 HID 幺五洞两，下一个点 HDE。

2. 全日空013，直飞 OLR VOR，保持当前高度层直到 OLR VOR，通过 HIG 275 度径向线报告。

3. 加拿大913，保持八千四，预计二十公里后上升高度。

4. 沈阳区域，四川7182，高度拐两保持，燃油续航能力只有四十五分钟，我们打算在沈阳加油。

5. 乌鲁木齐区域，南方3375，机翼前缘结满了冰，我们很难保持高度，申请立即下降。

汉译英

1. AFR 317, stop climb, converging traffic over DNM, maintain 9200 meters, report passing DNM.

2. Bravo Control, UAL 191, over SMK, ETO RK at 07, 10400 meters maintaining, request descent.

3. CXA 6198, Wuhan Control, maintain 550 kilometers per hour or less until passing WG.

4. PANPAN PANPAN PANPAN, Sierra Control, BAW 612, we've just intercepted an urgency call from AAL 096, a woman passenger is about to give birth, request divert to Sierra airport, the aircraft's position is 45 miles south of HGR, FL290.

5. Foxtrot Control, SAS 814, 40 miles west of BCP, FL 310, starboard engine flamed out, we are now trouble-shooting.

Lesson Ten

Ⅰ. Fill in the blanks.

1. estimating
2. controlled/maintain
3. over/estimating
4. negative/expect
5. lose
6. due
7. proceed/pattern

168

Key to exercises

8. join

9. available/accept

10. next

II. Listening practice.

1. enter/northwest/276/10/maintain/276

2. permission/leave/passing

3. engine/power

4. altitude/advised

5. boundary/maintaining

6. passing/LMG/1315

7. joining

8. severe turbulence/assistance

9. ACA 292, go ahead

ACA 292, cleared to B, flight planed route, FL276, join G12 at ROD

ACA 292, correct

10. DLH 721, Charlie Control, go ahead

DLH 721, cleared to cross Upper Red 207 at HGT, FL291

DLH 721, report HGT

III. Translation.

英译汉

1. 南方3392，保持拐八，在BEBEM加入航路Red两洞洞。

2. 南方3392，保持九千八，在五五分在翁源穿越航路Hotel两五，由于活动。

3. 东方2180，可以经KTD离开管制空域，在管制空域内保持高度层两九洞。

4. N231，MK以南十海里，航向两幺洞，通过九千英尺，预计PS三洞分，申请在PS穿越航路V538。

5. 国际1518，在VW等待，高度幺洞幺，预计穿越指令洞五分。

汉译英

1. Wuhan Control, CES 5321, boundary at GOSMA, ETO LKO 02, 10400m maintaining, request cross boundary.

2. CSN 3307, unable to approve 10100m due to opposite traffic at 9500m, estimated time of passing 13.

3. ANZ 538, B767 at 10400m, 80km to ZHO, request cross V21 at ZHO.

4. CAL 505, 10100m is not available due traffic, alternative is 10700m, advise.

5. JAL 781, hold at XS, maintain 5100m, traffic congestion at Baiyun airport, expect further clearance at 20.

169

Key to exercises

Lesson Eleven

Ⅱ. Fill in the blanks.

1. ready
2. estimating/information/report/levels
3. expedite/Leaving
4. maintain
5. cleared/track/pattern/further
6. hold
7. ready/altitude/Leaving/900ft on QNH 1005

Ⅲ. Listening practice.

1. cleared/PNK/4000feet/reaching
2. direct/maintain
3. over/5000 feet/31/outbound
4. present/standby
5. Commence/even
6. published/approach
7. ACA321, go ahead

ACA321, maintain FL210, expedite descent before 1835

8. AFR 626, hold at NCS at 9000ft, inbound track 120 degrees, left hand pattern, outbound time 1 minute, expect further clearance 50

9. SIA 717, hold as published, 7000ft due traffic congestion, expect approach time 1105

SIA 717 hold at LGS VOR/DME at 20 DME 7000ft, inbound track 210 degrees, left turns, outbound distance 24 DME

10. AFR 288, go ahead

AFR 288, descend to 6000ft on reaching MSK, hold as published, expect approach time 15

Ⅳ. Translation.

英译汉

1. 由于设备原因,不能保持 RVSM,恢复正常间隔。
2. 加拿大 505,武汉进近,可以飞往 WG 等待,保持高度幺两,入航航迹幺洞六,左航线,出航时间一分钟,预计进一步许可两三分。
3. 南方 3572,下降到四两,尽快下降通过四千八,过偶数高度层报告。
4. 国际 1578,在 BCH 上空等待,高度两千四,入航航迹两幺洞,左航线,出航时间一分钟,预计进近时间幺两分。
5. 日航 109,在 NBW VOR 一百七十度径向线距台二十五海里到三十五海里 DME 之间等待,高度七千英尺,入航航迹三五洞,右航线,预计十分钟后收到进一步许可。

Key to exercises

汉译英

1. CCA 1567, ADS-B transmitter appears to be inoperative, stop ADS-B transmission.

2. CCA 101, hold at VYK, 5400m, inbound track 360, left-hand pattern, outbound time one minute, expect approach time 15.

3. DAL 612, follow HZ−1A Arrival, descend and maintain 1200m, QNH 1013, make standard ILS approach runway 22.

4. CES 2829, cleared VOR approach runway 35, descend at own discretion, you are number one, report runway in sight.

5. JAL 741, stop descent at 6600m due outbound traffic.

Lesson Twelve

II. Fill in the blanks.

1. approach
2. maintain/expected
3. proceed/sight
4. approach/further
5. expedite
6. minimum
7. localizer
8. estimating
9. straight-in/visual/VOR⋯, established
10. altitude/heading

III. Listening practice.

1. visual approach/maintain
2. cleared/descend and maintain
3. intercept localizer/at your own discretion
4. unable
5. straight-in/descend/QNH 1018
6. ACA 541, go ahead

ACA 541, descend to 7000ft QNH1015, report HRM

ACA 541, cleared for ILS approach runway 17, you are number 3

7. DLH 709, pass your message

DLH 709, negative for VOR approach due traffic congestion

DLH 709, cleared localizer back course approach, runway 17

8. JAL 414, Charlie Approach, cleared NDB approach runway 32, descend to 900m

JAL 414, negative due to traffic on ground, continue NDB approach

171

Key to exercises

IV. Translation.

英译汉

1. 国际1579,预计目视进近,跑道三六右,直飞QU,下降到幺八保持,目视跑道报告。
2. 东方5321,继续进近,我将与你保持联系。
3. 南方3308,武汉进近,收到MAYDAY,最近合适的机场是武汉天河机场,可以直飞DA,下降到两拐保持。我们将通知武汉塔台准备你落地。
4. 达美901,可以VOR进近,跑道幺八右,下降自己掌握,你是第一个落地,目视跑道报告。
5. 东方5402,武汉进近,可以飞往WG,加入标准等待程序,预计进一步许可两洞分。

汉译英

1. No. 2 engine intensive vibration, reason unknown, we might shut down the engine anytime, request approach immediately.
2. CES 5301, Wuhan Approach, follow R10D RNAV departure.
3. CSN 3342, Guanghan Approach, cleared for straight-in NDB approach runway 22.
4. Hold on the 300 radial of RAL VOR between 25 miles and 30 miles DME at 3000m, inbound track 120, left hand pattern, delay not determined.
5. CES 5301, Wuhan Control, unable RNAV, radar vector direct to WUH VOR.

Lesson Thirteen

I. Fill in the blanks.

1. follow
2. straight-in
3. extend/final/Roger/sight
4. orbit/report
5. maneuver/spacing/preceding
6. join
7. field/visual
8. circuit/joining
9. orbit/final
10. continue/outer maker

II. Listening practice.

1. short/cleared/surface wind
2. short approach/G-CD
3. traffic/taking off
4. traffic/clear/downwind
5. 25km/landing
6. hold short/17L/Break Break/one orbit right/further

Key to exercises

7. right base/270,6 knots

8. CSN 3561, number two, follow the B737 on final, CSN 3561, base

9. CES 2910, number 1 to land, wind calm

Going around. confirm the standard procedure, CES 2910

10. JAL 556, descend to circuit height 600m, QNH 1012, join right hand pattern, runway 25

JAL 556, report base, number 2, B757 6 miles on final

JAL 556, turning base

JAL 556, continue approach, wind 220, 10 knots

III. Translation.

英译汉

1. 南方3301，你是第二个落地，第一架飞机正在接地，过外指报告。
2. 加拿大185，当前位置左转，盘旋着陆跑道洞四，你是第一个落地。
3. 国际1834，加入右三边，跑道幺五，地面风幺拐洞，十节，修正海压幺洞洞幺。
4. 法航501，左右做机动因为间隔，你离前机太近。
5. 大韩889，下降到起落航线高度九百，场压九九八。

汉译英

1. CES 5230, extend downwind, number two, B737 5km final.
2. BAW 857, make one orbit right due traffic on the runway, report again on final.
3. Fuzhou Tower, we are unable to align with the center line at decision height, possibly due gusting wind, now everything is normal, CSN 3245.
4. JAL123, extend downwind 2km to give room for a departure, report turning base.
5. CSN 3385, Tower, go around as published, standby for further instruction.

Lesson Fourteen

II. Fill in the blanks.

1. go around
2. loss/windshear
3. closed
4. hold/diversion
5. outer/advise/make/cleared
6. touch and go/circuit
7. land/final/Going around/procedure

III. Listening practice.

1. missed approach
2. full stop, negative for touch and go/congestion/180 degrees 5knots
3. expected approach time/at minimum of 45/unable/30 minutes/diverting

Key to exercises

4. low/training

5. extend

6. final/traffic B737 taking off and an A320 downwind/vacate/Going around, confirm standard procedure/climb straight ahead/119. 5

IV. Translation.

英译汉

1. 东方2827,跑道有障碍物,立即复飞,我重复一遍,复飞。

2. 南方3210,不能做落地连续,因为交通拥挤,做全停,跑道两幺,可以落地。

3. 加拿大108,沈阳进近,机场关闭,由于正在扫雪,可以备降长春。

4. 国际1509,遭遇严重晴空颠簸,机上有旅客受伤,申请备降到离我们最近的美兰机场。

5. 塔台,B2890,看到跑道了,申请落地连续做训练。

汉译英

1. Wuhan Tower, JAL 223, we are going around, we can't extend our landing gear due hydraulic system failure.

2. Beijing Tower, BAW 411, left engine broke down due birdingestion, we are returning, request priority landing.

3. Xiamen Tower, CXA 8837, no contact at minima, pulling up.

4. Nanjing Tower, CHH 7374, we can't crank landing gear down, we'd like to go around and direct to the holding area for check.

5. Hong Kong Tower, JAL 520, request low pass for visual landing gear check.

Lesson Fifteen

I. Fill in the blanks.

1. heading/contact

2. maintain or descend/vectoring

3. heading/intercept/established

4. take/contact/vacated

5. time/backtrack

6. final/surface/cleared

7. outer/approach/cleared to/take/vacated

II. Listening practice.

1. glide path/170 knots/121. 8

2. taxiway B/F intersection/parking/marshaller in sight

3. 10 miles/present speed 180 knots/118. 7

4. Ground/vacated/left/B4/P2/stand 30/Via

174

Key to exercises

5. first right/121. 5/runway vacated/second left/Delta7

6. cleared to land/weare unable to vacate runwag/we need their help

7. AFR 901, cleared to land, wind 150 degrees 10knots

AFR 901, go around, a vehicle crossing the runway

AFR 901, climb and maintain 1200m, join right downwind, runway 15

Ⅲ. Translation.

英译汉

1. 北京地面,印度182,我们刚刚冲出滑行道 Delta,左主轮似乎陷在泥里,能否为我们派一辆拖车?

2. 法航671,前方第二个道口右转,脱离后联系地面么么九点两。

3. 北京塔台,荷兰057,我们不熟悉本场,请派引导车将我们引导到停机位。

4. 南方3109,注意:侧风两八洞,八节,可以落地。

5. 贵阳进近,奥凯8509。近地告警,拉起来了,上升到么五,QNH 么洞么三。

汉译英

1. JAL 850, landing time 45, backtrack runway 21.

2. CES 5820, vacate runway next right, taxi to Gate 04 via taxiway B4, D1 and C1.

3. CCA 1718, caution, preceding aircraft reported wind shear along your landing course, cleared to land runway 18L.

4. Beijing Tower, CES 2837, request gear up landing and foam carpet on the runway.

5. DLH 917, Xi'an Tower, we have informed the medical department (first-aid service), doctor and ambulance are to meet you at the terminal apron.

Lesson Sixteen

Ⅱ. Listening practice.

1. identification/070

2. left/spacing

3. resume/southwest

4. one orbit/delaying action

5. squawk normal

6. own navigation/magnetic

7. separation/lose/or later

8. traffic congestion/right

9. level/4900/for identification turn left heading 020/heading 020/identified/airfield

10. 8400/for identification turn right heading 300/heading 300/observed on radar/Turning back to 270, resuming own navigation

Ⅲ. Translation.

汉译英

1. CCA 1823, for identification, turn left heading 230.

Key to exercises

2. CES 2587, radar contact, 8 kilometers southwest of VMB, continue present heading.

3. CSN 3357, radar service terminated, resume own navigation, direct to VYK.

4. CCA 1508, traffic information, unknown traffic, 12 o'clock, 5km, opposite direction, fast moving.

5. CSN 3128, clear of traffic, resume own navigation, direct to VYK.

英译汉

1. 国际1326,雷达已识别,省略位置报告直到VMB。

2. 南方3308,为了延迟,左转一圈。

3. 国际1307,恢复自主领航,直飞CH,航迹洞三洞,距离二十公里。

4. 南方3306,活动通报。不明活动,十点钟方位,十海里,从左向右穿越,快速移动。

5. 南方3103,右转,航向洞四五,可以仪表进近,跑道幺三,建立盲降报告。

Lesson Seventeen

II. Fill in the blanks.

1. terminated
2. proceed
3. available
4. failure
5. passed clear
6. squawk
7. hold
8. o'clock
9. crossing
10. navigation

III. Listening practice.

1. negative/vectors
2. avoiding action
3. 090/6
4. floating object/proceed with caution
5. closing/negative contact/Leaving/clear of traffic

IV. Translation.

英译汉

1. 日航021,浦东塔台,机场西南发现不明飞行物,慢速移动,小心飞行。

2. 国际1508,尽快下降到八千四,有飞机穿越。

3. 南方3379,雷达失效,转换到非雷达间隔,左右做机动增大间隔。

4. 立荣102,现在雷达工作电力不足,看不到一次雷达回波,为了在三五分过HG,调整你的速度。

5. 四川9755,雷达失效,我们不能再协助你了,为增大安全间隔,速度减少三十节,过

Key to exercises

SWK 报告。

汉译英

1. All stations, SSR malfunctioned, PSR signal is unstable, maintain own separation and proceed with caution, repair is under way.

2. Nanjing Approach, CSN 3125, our ADF on board is out of order, we have trouble maintaining the right course, request emergency assistance.

3. ANA 988, the restricted area 65km ahead of you is activated, turn right 35 degrees immediately to avoid it.

4. CSN 3509, Xi'an Approach, traffic information, traffic 9 o'clock, 7km, overtaking, A320, 3600m.

5. Radar failure, report endurance.

Lesson Eighteen

II. Translation.

汉译英

1. CSN 3281, weather is deteriorating due sandstorm, RVR runway 13 is less than 800m, advise executing missed approach.

2. Shenzhen Tower, CES 5697, the weather is below captain minimum due to strong precipitation, we would like to delay our approach time until the weather at Bao'an airport improves.

3. CSH 2896, Wuhan Tower, the weather is rapidly deteriorating, fog is coming down and obscuring the upwind end of runway 06, visibility less than 700m, please taxi back to the parking stand, I will keep you advised.

4. All stations, Wuhan Tower, current ceiling is 300m, expected to rise to 650m within the next 25 minutes, airport will be reopened shortly.

5. CCA 1472, ceiling 300 meters, variable, visibility 2000m in fog, wind 250 degree, 3m/s.

英译汉

1. 郑州区域,大韩 866,发动机马力不足,由于发动机进气受阻,机翼前缘结满了冰,怀疑进气道也被冰覆盖,申请立即下降。

2. 广州区域,南方 3309,现在脱离了结冰区,正在返回航路。

3. 北京区域,东方 5109,偏航阻尼失效,在我们的前方有严重结冰吗?

4. 新航 195,南京塔台,天气正在恶化,跑道洞六 RVR 小于一千五百米,报告意图。

5. 南方 3392,在停机位 Charlie 五等待,我们已经通知相关部门,他们会立刻给你派除冰车。

III. Listening practice.

1. runway 36L/850m/not availabk/700m

177

2. visibility/RVR/less than

3. experiencing/icing/weather

4. severe icing/level/de-icing/emergency

5. continuous severe icing

6. severe icing area/advise

7. lower visibility

8. de-icing system/iced over/descent

9. experiencing/climb/climb to 9500m

10. de-icing/resumed/report reaching

Lesson Nineteen

Ⅱ. Listening practice.

1. 50km/go round it

2. rear fuselage damaged/priority landing and radar vector

3. severe turbulence/divert to Hefei

4. check the type of traffic that's ahead of us

5. 30 degrees 40km offset left

6. by turning 30 degrees left track out 25km

7. rapidly approaching the far end of the runway

8. circumnavigate the build-up at your own discretion

9. climb to 11000m to overfly the CB

10. call when ready to leave my frequency

Ⅲ. **Translation**.

英译汉

1. 南京进近,南方3097,跑道两四五边五公里遭遇强风切变,高度两百米,差点失速。

2. 北京区域,东方5109,偏航阻尼失效,在我们前方有颠簸吗?

3. 武汉区域,海南8516,脱离雷暴,正在返回航路。

4. 广州区域,快达191,申请备份航路,因为我们上次遭遇颠簸,结果两台发动机都顺桨了。

5. 广州区域,上航405,发动机重新启动,看来工作很稳定,刚才碰到上升气流时熄火了。

汉译英

1. Xi'an Control, JAL 295, we have an indication of weather 40km ahead of us on our present heading, request go round the CB from the left.

2. Zhuhai Approach, CHH 7153, we have encountered lightning strike on take-off, we are having trouble controlling the pitch, elevator probably damaged, request return for landing immediately.

Key to exercises

3. Shanghai Control, CSZ 9175, level at 7200m, encountering severe clear air turbulence, passengers are injured, request level change to avoid the turbulent area.

4. Lanzhou Control, KAL 866, engine is low on power due engine starvation, our leading edge is iced over, suspected air inlet also covered with ice, request immediate descent.

5. CXA 8126, Beijing Control, thunderstorm 45km ahead of you, circumnavigate the CB by turning left 30 degrees, track out 25km from your present position.

Lesson Twenty

Ⅲ. Listening practice.

1. aborted/executing/evacuation/ambulance
2. dump fuel/limitations/jettison/completion
3. receiver/transmission
4. extinguished/fire service assistance
5. ingested/flamed out
6. hijackers/hijack/intimidated
7. radio contact/observed
8. windshield/confirm
9. smashed/strike
10. foam/dumping
11. gear up/foam carpet/emergency/foamed

Ⅳ. Translation.

英译汉

1. 广州区域，美联航198，我们很难保持航迹，飞机有右偏的趋势，方向舵可能失灵了。

2. 广州进近，国际1898，过热灯闪亮了，滑油需要冷却。

3. 紧急！上海区域，南方3509，三发强烈抖动，原因不明，我们可能随时关闭发动机。

4. 紧急！成都区域，日航105，四号发动机整流罩冒出浓烟，可能着火。

5. 成都进近，国泰508，外层风挡好像被冰雹或闪电击碎，因为我们听到很响的重击声。

汉译英

1. PANPAN PANPAN PANPAN, Harbin Control, CSH 5182, our altimeter is out of service, request lower level to get below the clouds.

2. Shenzhen Tower, CDG 8291, brakes and steering inoperative due to hydraulic trouble.

3. Dalian Control, SAS 583, hydraulic system is still leaking, but we'll be able to manage it to Beijing.

4. PANPAN PANPAN PANPAN, Guangzhou Approach, CSZ 8829, intercepted urgency call from CES 2837, they are getting low on fuel, request priority landing, position 10km northwest of the field at 3000m.

Key to exercises

5. Guangzhou Approach, CCA 1307, we request a doctor on arrival at Shenzhen because one of the crew members is not feelingwell, the message to be copied to Air China.

Lesson Twenty-one

II. Fill in the blanks.

1. fire/oxygen/discretion/priority/straight-in
2. struck/pulled or sucked/transmitting/emergency/level

III. Listening practice.

1. leak/downwind/left or right
2. 5 miles/034/vector to IGS
3. power off/control/security assistance
4. squawking/ident/approach clearance desired
5. hydraulic/main gear/manual

IV. Translation.

英译汉

1. 上海区域,国际1279,我们被劫持了,这家伙要我们改航马尼拉降落,但我们需要在浦东机场加油,申请立即下降。

2. MAYDAY,MAYDAY,MAYDAY,北京进近,南方3158,机上有炸弹爆炸了,客舱起火,飞机正在解体,迫降在大王庄以南二十海里。

3. 大连区域,新航311,左发在上升过程中吸入飞鸟,发动机熄火。

4. 全体注意,VY和WX之间有紧急下降的航空器,VY和WX之间所有低于高度三千的航空器立即向东离开走廊。

5. 成都进近,东方2813,我们油量不够了,由于流量控制,我们已经等待了一个小时,申请在成都双流机场立即着陆。

汉译英

1. MAYDAY, MAYDAY, MAYDAY, Beijing Approach, CSN 3309, number two engine flamed out due unknown foreign object strike, suspected bird ingestion, request priority landing and emergency services on arrival.

2. Jinan Control, ANA 839, we are climbing, part cowling of port engine shed, hit the tailplane, we are having trouble controlling the pitch, request come back and priority landing.

3. Beijing Control, DLH 734, cancel distress, we have resumed control of the aircraft, we'd like to divert to Zhengzhou.

4. MAYDAY MAYDAY MAYDAY, Wuhan Approach, UIA178, we've just been hit by a large vulture on climbing out, first officer windshield cracked, we are in danger of decompression, advise.

5. Shenyang Approach, CSN 3085, we intend to return to land, but because of landing weight requirements dictated by structural limitations of the aircraft, we have to dump 20 tons of fuel first.

参考文献(References)

[1] O'DOHERTY P, CONSIDINE B. Guidelines for Controller Training in the Handling of Unusual/Emergency Situations[M]. 2003.
[2] NATS Aircraft Emergencies-Considerations for Controllers[M]. UK: The Stationery Office, 2005.
[3] International Civil Aviation Organization. Manual on the Implementation of ICAO Language Proficiency Requirements[M]. 2010.
[4] 潘卫军,肖靖. 空中交通无线电通话用语指南[M]. 成都:西南交通大学出版社,2005.
[5] 吴土星. 陆空通话教程[M]. 北京:中国民航出版社,北京,1996.
[6] LEVESON L F, CASS M. Sky talk[J]. Air Transport World, 1997 1,4(11):32.
[7] ROBERTSON F A, JOHNSON E. Airspeak[M]. USA: Prentice Hall College Div, 1990.
[8] RENGAD P Y, POTTER C, ROVES G. English for pilots and controllers[M]. France: Ecole Nationale de l'Aviation Civile, 1988.
[9] 中华人民共和国交通运输部,2017 年第 30 号,《民用航空空中交通管理规则》(CCAR-93TM-R6). 2017-09-29.
[10] 刘继新. 特殊情况下的无线电通话用语[M]. 北京:国防工业出版社,2010.
[11] International Civil Aviation Organization. Manual on the Implementation of ICAO Language Proficiency Requirements[M]. 2007.
[12] 美国联邦调查局. 民用航空术语汇编[M]. 中国民用航空总局航空器适航审定司,中国民用航空总局航空安全技术中心,译. 2001.
[13] 中国民用航空局空中交通管理局. 空中交通无线电通话用语[M]. 2022.
[14] 中国民用航空局空中交通管理局. 空中交通无线电模拟通话手册[M]. 2022.